EXPLANATORY NOTES

The views expressed in this publication are those of the author and do not necessarily reflect the views of the United Nations Industrial Development Organization (UNIDO).

Besides the common abbreviations, symbols and terms, the following have been used in this publication:

Economic and technical abbreviations

BL	biosafety levels
DNA	deoxyribonucleic acid
rDNA	recombinant deoxyribonucleic acid
GPMs	good manufacturing practices
PMN	pre-manufacturing notification
RNA	ribonucleic acid
rRNA	recombinant ribonucleic acid

Organizations

ACGM	Advisory Committee on Genetic Manipulation (United Kingdom)
BSB	Biotechnology Science Board (United States)
BSCC	Biotechnology Science Coordinating Committee (United States)
CDC	Centres for Disease Control (United States)
EFB	European Federation of Biotechnology
EPA	Environmental Protection Agency (United States)
FCCST	Federal Coordinating Council for Science and Technology (United States)
FDA	Food and Drug Administration (United States)
FIFRA	Federal Insecticide, Fungicide, and Rodenticide Act (United States)
GMAG	Genetic Manipulation Advisory Group (United Kingdom)
HSE	Health and Safety Executive (United Kingdom)
IBC	Institutional Biosafety Committee
NEPA	National Environmental Policy Act (United States)
NIH	National Institutes of Health (United States)
NIOSH	National Institute for Occupational Safety and Health (United States)
NSF	National Science Foundation (United States)
OECD	Organization for Economic Co-operation and Development
ORDA	Office of Recombinant DNA Activities (of NIH)
OSHA	Occupational Safety and Health Administration (United States)
OSTP	Office of Science and Technology Policy (United States)
PWG	Plant Working Group (of RAC) (United States)
RAC	Recombinant DNA Advisory Committee (of NIH)
TSCA	Toxic Substances Control Act (United States)
USDA	United States Department of Agriculture

CONTENTS

Annexes

Introduction

Old and new biotechnology

Biotechnology, the use of living organisms for industrial, agricultural or medical purposes, is not new. For thousands of years, people have used plants and animals to provide food, materials and medicines and have used micro-organisms to process sugar to alcohol or milk to cheese. Micro-organisms have been used to prepare vaccines for over a 100 years and to prepare antibiotics for nearly 60 years.

The genetic manipulation of living organisms is also not new. For centuries, farmers and animal breeders have engaged in a crude form of genetic manipulation by continuously selecting and mating those plants and animals that had desired characteristics. Scientists have used similar selection technologies to develop high-yielding strains of antibiotic-producing micro-organisms and have engaged in other genetic manipulations, such as mutating micro-organisms using chemicals or X-rays.

In the last 20 years, however, powerful new techniques have been developed that greatly increase the ability of scientists to manipulate the inherited characteristics of plants, animals and micro-organisms. These techniques, which may be generally referred to as genetic engineering, include the direct chemical synthesis of genes and proteins, recombinant deoxyribonucleic acid (rDNA), recombinant ribonucleic acid (rRNA), cell fusion, plasmid transfer, transformation, transfection and transduction.[1] In this publication, these genetic engineering techniques and the use of organisms produced by them will be called the new biotechnology, in order to distinguish them from traditional breeding and fermentation techniques.

The new biotechnology promises to have a profound impact upon the human condition. It may contribute to filling some of the most fundamental needs of humanity, from health care to supplies of food and energy to pollution control. Potential applications of the new biotechnology include the production of new drugs, food and chemicals, the more efficient production of existing products, new diagnostic techniques, the degradation of toxic wastes and major improvements in agricultural products.

Along with its promise, the new biotechnology has raised concerns about possible risks to humans, animals and the environment. It might be asked whether genetically engineered organisms could be harmful to humans or other living organisms; whether, if some of these organisms establish themselves in the environment, they could proliferate and become pests; and whether some of the powerful new chemicals that can now be manufactured by the new biotechnology could prove harmful to factory workers.

---

[1]See the glossary for a definition of terms.

International impacts

The impact of genetic engineering will be international in scope. Accordingly, it is not surprising that the United Nations and its affiliated organizations have made a major commitment to become parties to the development and uses of biotechnology. A desire for the developing countries to share in the benefits of biotechnology sparked efforts by the United Nations Industrial Development Organization (UNIDO) to create an international centre to promote the development and peaceful application of genetic engineering and biotechnology, especially for developing countries. The centre, known as the International Centre for Genetic Engineering and Biotechnology (ICGEB) is supported by 41 countries and operates at Trieste, Italy, and New Delhi, India, under the auspices of UNIDO.[2] Concern over possible safety and environmental risks raised by biotechnology has prompted the World Health Organization (WHO) and the United Nations Environment Programme (UNEP) to identify and study the various safety issues involved. WHO has long been concerned about the risks to human health presented by mirco-organisms and has prepared a laboratory biosafety manual.[3] The WHO Regional Office for Europe has prepared a report on the health impacts of biotechnology.[4] UNEP has been studying the topics of biowaste disposal and environmental uses of genetically engineered organisms. These activities have led to an affiliation between UNIDO, WHO and UNEP in the form of an informal working group to exchange information and co-ordinate activities pertaining to safety issues raised by the new biotechnology.

This publication was originally presented as a paper to a meeting of the informal working group, held at Vienna from 27 to 29 January 1986. It was written by Geoffrey M. Karny, UNIDO consultant.

Purpose and structure of the publication

One of the many topics that ICGEB will need to address will be the safety issues raised by the new biotechnology. Moreover, as an international centre of excellence, presently supported by 41 countries with more expected to join in the future, the Centre will naturally be looked upon as a leader and an international model for dealing with biotechnology, including issues of safety.[5] Therefore, it is appropriate to identify those issues and to consider what role international bodies such as WHO, UNEP, UNIDO and ICGEB can play in addressing them. The purpose of this publication is therefore:

(a) To examine current views on the possible risks presented by the new biotechnology;

(b) To identify any safety issues or concerns arising from such risks;

---

[2]Statutes of the International Centre for Genetic Engineering and Biotechnology (ID/WG.397/8) article 21. The Centre will become independent when at least 24 countries have ratified its statutes.

[3]*Laboratory Biosafety Manual* (Geneva, World Health Organization, 1983).

[4]*Health Impact of Biotechnology: Report on a WHO Working Group, Dublin, 9-12 November 1982* (Copenhagen, World Health Organization, 1984).

[5]At present, the safety guidelines of the United States National Institutes of Health are being observed at the Centre. The Centre is also monitoring the debate concerning physical/biological containment as well as the release of genetically engineered organisms into the environment.

(c) To examine current international regulatory and supervisory mechanisms for dealing with the risks and safety concerns;

(d) To determine common approaches and identify any potential gaps, overlaps or other deficiencies in these mechanisms;

(e) To discuss the international significance of these matters;

(f) To propose roles for ICGEB, WHO, UNEP and UNIDO.

In addressing these issues, the author has drawn upon other published reports and interviews with selected experts[6] as well as personal experience. An attempt has been made to draw on all available written materials pertaining to risk but only on selected materials pertaining to regulation. It would be unnecessary and virtually impossible to survey the laws of many different countries with regard to such broad areas as worker health and environmental protection. Thus, only a few bodies of law were selected for an in-depth study. They were the laws of Japan, the United Kingdom of Great Britain and Northern Ireland, the United States of America, and the European Economic Community. The three major industrial countries have well-developed, comprehensive laws from which common regulatory principles may be ascertained. Thus, such a study is directly relevant to the interests of less developed countries, which can select from and build upon these principles.

Even this approach has limitations, however. There is always difficulty in determining and interpreting the laws of countries other than one's own. Material may not be readily available in written form; a substantial body of law may exist only in the form of unwritten policies and practices. Nevertheless, it is believed that sufficient information can be gathered from which to draw common principles that can be applied to the issue of how to regulate the risks presented by genetic engineering.

The new biotechnology will have many different commercial and scientific applications. For purposes of discussing risk and regulation, however, it will be most useful to group the applications into three categories: laboratory-scale research; large-scale industrial processes; and environmental uses of genetically modified organisms.

---

[6]Extensive interviews were conducted with the following experts: J. Ian Waddington, Director, Environmental Health Service, WHO Regional Office for Europe; Jorma O. Jarvisalo, Regional Officer for Occupational Health, WHO Regional Office for Europe; Michael Suess, Regional Officer for Environmental Health, WHO Regional Office for Europe; Vinson R. Oviatt, Director, WHO Programme on Safety Issues in Microbiology; Hamdallah Zedan, United Nations Environment Programme; and Lech J. Pierkarski, United Nations Environment Programme.

I. Risks and regulation of laboratory research

Early concerns about risks

In the early and mid-1970s, molecular biologists developed the powerful new technique of gene splicing or DNA. This technique allowed scientists to take the DNA from one organism and place it in another organism in such a way that it continued to function and produce its normal product. In other words, the second organism expressed the trait controlled by the product of the gene of the first organism. Although the state of scientific knowledge was such that the ability to practice the technique was essentially limited to micro-organisms, it was apparent that this technique and others could someday be applied to the cells of higher organisms, such as plants, animals and humans.

The power of the new technique and the recognition of how little was actually known about the genetics and physiology of living organisms caused many of the molecular biologists to wonder if dangerous new organisms might be created inadvertently. In an unprecedented move, a committee of prominent scientists involved in rDNA research called for a temporary world-wide moratorium on certain types of experiments in July 1974 and also called for an international conference on potential biohazards of the research.[7] The scientists also requested that the director of the United States National Institutes of Health (NIH) should consider establishing an advisory committee to develop a programme to evaluate potential hazards and to establish guidelines for experimenters. In response, the director of NIH established the Recombinant DNA Advisory Committee (RAC) on 7 October 1974, and an international conference was held at the Asilomar Conference Center, Pacific Grove, California, in February 1975. The Conference concluded that, although a moratorium should continue on some experiments, most work involving rDNA could continue with appropriate safeguards in the form of physical and biological containment.[8]

The ensuing work of the RAC, many scientists, public health professionals, lawyers and interested members of the public culminated in the NIH guidelines for research involving recombinant DNA molecules promulgated in June 1976.[9] The NIH guidelines prohibited certain experiments and required the use of containment procedures and other safeguards for other types of experiments. In addition, the guidelines created an oversight mechanism designed to assign responsibility to the individuals and institutions involved in research and to

---

[7]*Impacts of Applied Genetics: Micro-Organisms, Plants, and Animals* (Washington, D.C., United States Office of Technology Assessment, 1981) (OTA-HR-132), appendix IIIA, p. 315.

[8]*Ibid.*, p. 316.

[9]United States of America, 41 Fed. Reg. 27,902 (1976).

monitor those parties for compliance with the guidelines. Although the guidelines technically applied to scientists and institutions receiving grants from NIH for rDNA research, other federal agencies soon began to require the compliance of scientists and institutions receiving grants. Industrial scientists espoused voluntary compliance. The guidelines were seen as fairly restrictive, and their adoption forestalled efforts in the United States Congress to pass legislation to regulate rDNA research. Similar guidelines were adopted in many other countries.[10]

As time passed, it became clear that the initial fears about the possible risks of rDNA research were greatly overstated. Several risk-assessment studies led to a downward evaluation of the potential risks. Knowledge gained from experience with the technique allayed much of the fear of probing the unknown. No evidence was brought forth to support many of the early risk scenarios, and, most significantly, there has been no evidence of any harm to human or animal health or the environment from rDNA. Finally, the input of experts in infectious diseases made it clear that more than the modification of just one or two genes was necessary to create a pathogenic organism and that such a modification generally would be deleterious to the survival of the organism. Consequently, the requirements of the rDNA guidelines in the United States have been substantially relaxed.[11]

Current views on risk

The current consensus among experts appears to be that rDNA techniques present no special risks beyond those inherent in the materials being used. In its recent report on commercial biotechnology, the United States Congressional Office of Technology Assessment concluded that "today, most experts believe that the potential risks of rDNA research were drastically overstated and that rDNA technology generally does not involve a risk beyond that already inherent in the host, vector, DNA, solvents, and physical apparatus being used".[12] Similarly, the World Health Organization concluded that "there are no unique or specific safety risks associated with recombinant DNA work (genetic engineering); the risks are no greater than those associated with work with known pathogens and do not necessitate special laboratory design or practice".[13] A WHO working group, however, while generally agreeing with these conclusions, cautioned that care should be taken before fully adopting this principle for all possible experiments and suggested that the judgement of experts should be sought for the review of processes employing DNA coding for highly potent toxins.[14]

---

[10]This information is based upon international surveys undertaken by the Committee on Genetic Experimentation of the International Council of Scientific Unions, reported as of July 1979. It is believed that this information is still essentially current. The 27 countries were: Australia, Belgium, Brazil, Bulgaria, Canada, China (Taiwan Province), Czechoslovakia, Denmark, Finland, France, German Democratic Republic, Germany, Federal Republic of, Hungary, Israel, Japan, Mexico, Netherlands, New Zealand, Norway, Poland, South Africa, Sweden, Switzerland, Union of Soviet Socialist Republics, United Kingdom, United States and Yugoslavia.

[11]*Commercial Biotechnology: An International Analysis* (Washington, D.C., United States Congress, Office of Technology Assessment, 1984) (OTA-BA-218).

[12]*Ibid.*, p. 355.

[13]*Laboratory Biosafety Manual, . . .*, p. 30.

[14]*Health Impact of Biotechnology . . .*, p. 25.

The view that rDNA techniques present no special risks beyond those inherent in the materials being used was generally concurred with by the experts interviewed for this publication. Hamdallah Zedan, however, expressed concern that there might be some special risk involved with rDNA if the technique caused the expression of an otherwise unexpressed gene that made a harmful product. This might occur if the foreign DNA disrupted a control that kept the harmful gene from being expressed. An example of such a gene would be an oncogene.[15] At this point, this is a conjectural risk. There has been no evidence to date of harmful products being inadvertently made by the rDNA technique.[16] The issue is being studied and would appear to be an appropriate area for future study by ICGEB.

With respect to other types of genetic engineering, the consensus of the experts also appears to be that they present little, if any, risk beyond that inherent in the materials being used. This consensus is not based upon specific risk assessment of these techniques. Rather, it appears to be based upon: the general experience of the scientific community; the recognition that these techniques were less powerful than rDNA; and greater experience with the techniques.

The concern about potential risks presented by genetic engineering has always focused on rDNA. It has been noted that the NIH guidelines did not address the full scope of risks of genetic engineering because they covered only recombinant DNA.[17] The apparent consensus of the scientific community, however, was that these other techniques did not merit the special attention and guidelines that rDNA merited. In view of the fact that there has been no reported incidence of harm arising from any of these other techniques and in view of the fact that there are few significant restrictions on laboratory research involving rDNA, this view appears to be reasonable at the present time.

The scientific community has come to view rDNA and the other genetic engineering techniques as part of a much larger collection of techniques used in the laboratory. As such, they should be governed by the existing framework of good laboratory safety practices.[18] The continuing existence of special guidelines for rDNA research, however, indicates that there are some special risks and uncertainties in the research that require special attention. On the other hand, the existence of such guidelines may simply reflect certain political realities—that the public continues to have lingering concerns about the safety of rDNA research and, therefore, requires assurance that there is some degree of supervision. Since general laboratory safety and specific rDNA guidelines are relevant to the issue of regulation of genetic engineering, both will be considered.

---

[15]Interview with Hamdallah Zedan, United Nations Environment Programme, Nairobi, Kenya (13 September 1985).

[16]Telephone interview with Elizabeth Milewski, Office of Recombinant DNA Activities, National Institutes of Health, Bethesda, Maryland, 3 January 1986. In her opinion, the inadvertent activation of oncogenes might be a concern in human gene therapy but probably would not be a concern in the use of genetically engineered micro-organisms to produce pharmaceuticals, chemicals and similar products in contained systems.

[17]*Impacts . . .*, p. 217.

[18]*Laboratory Biosafety Manual, . . .*, p. 30.

Regulation of genetic engineering in the laboratory

Many countries, in particular Japan, the United States and the western European countries, have a multitude of laws and regulations designed to protect the environment and the health and safety of workers and the public. Few of these appear to have been drafted specifically to cover laboratory research, however. There are several reasons that could explain this situation. First, there may be some question of whether there is sufficient authority under the statutes to regulate laboratory research or at least something as specific as genetic engineering. When the United States Government was wrestling with the question of how to address the perceived risks of rDNA research, a federal inter-agency committee concluded that there was not sufficient statutory authority to regulate such research in the manner and to the degree thought desirable at the time.[19] On the other hand, the United Kingdom's guidelines covering rDNA research were promulgated pursuant to statutory authority, the Health and Safety at Work etc. Act 1974.[20] Secondly, while certain kinds of laboratories, particularly those handling infectious micro-organisms, have been seen as presenting risks, the risks were seen as being mainly to those who were working with the agents rather than to the community at large.[21] Those at risk were presumably highly trained individuals who understood the risks and were capable of taking appropriate safeguards. Thirdly, various written and unwritten good laboratory practices have been developed by experts and apparently have been sufficient to protect against any hazards in the laboratory.

Whatever the reasons, the existing regulation of genetic engineering in the laboratory primarily consists of written and unwritten good laboratory practices, some of which are specifically directed to rDNA research. These practices or guidelines are essentially a self-regulatory mechanism created by scientists and public health specialists.

The only guidelines specifically directed towards genetic engineering in the laboratory are guidelines for the conduct of rDNA research. Such guidelines have been adopted by approximately 27 countries[22] and are generally similar in their approach. In fact, most of the countries that adopted guidelines modelled them after either the United Kingdom or the United States, although Canada, France, the Federal Republic of Germany and the Soviet Union drafted their own versions.[23]

These guidelines are based upon certain assumptions and principles. First, organisms containing rDNA may have to be contained. Secondly, the level of containment should be related to the degree of perceived risk. Thirdly, there should be some oversight of rDNA work. Fourthly, the degree of oversight

---

[19]*Interim Report of the Federal Interagency Committee on Recombinant DNA Research: Suggested Elements for Legislation* (Washington, D.C., 15 March 1977, pp. 9-10.

[20]*Commercial Biotechnology . . .*, p. 553.

[21]*Biosafety in Microbiological and Biomedical Laboratories* (Washington, D.C., United States Department of Health and Human Services, Public Health Service, Centers for Disease Control and the National Institutes of Health, 1984), HHS Publication No. (CDC) 84-8395, p. 2.

[22]See note 10.

[23]*Impacts . . .*, p. 322.

should increase with the degree of perceived risk.[24] The implementation of these concepts is fairly similar in the various countries because the world-wide scientific community was involved in their development and because most countries followed the lead of NIH. There are some important differences in the guidelines as adopted in the various countries, however, and different countries are at different stages in the process of relaxing them.[24]

In view of the leading role that the NIH guidelines have played as a model for other countries, it is appropriate to examine them first and in detail. In addition, the guidelines of Japan, the United Kingdom and the European Economic Community will also be examined.

NIH rDNA guidelines

The NIH guidelines are primarily oriented towards laboratory-scale research, although they contain provisions regarding the physical containment of large-scale work (work involving more than 10 litres of culture) and certain uses of plants containing rDNA in the environment. (These latter provisions will be discussed in chapters II and III of this publication.) The guidelines were originally promulgated in June 1976 and have been revised several times since then to substantially relax their restrictions. The latest complete version was published on 23 November 1984;[25] a copy is attached as annex I.

The NIH guidelines apply to all rDNA research[26] in the United States and its territories conducted at or sponsored by any institution receiving support for rDNA research from the federal Government,[27] including federal laboratories. Compliance is enforced by the authority of the federal Government to suspend, terminate or place restrictions upon its financing of the offending project or all rDNA projects at the institution receiving support.[28]

Although the NIH guidelines are not legally binding upon private companies (unless the company receives federal funds), the private sector has espoused voluntary compliance.[29] Moreover, some states and localities have required industry to comply by law. There has been no evidence that private companies in the United States have not followed the guidelines.[30]

---

[24]See, in general, *Commercial Biotechnology . . .*, p. 357.

[25]United States of America, "Guidelines for research involving recombinant DNA molecules", 49 Fed. Reg. 46,266-91 (1984).

[26]*Ibid.,* 46,267, sect. I-B; research involving the construction and handling of rDNA molecules and organisms and viruses containing them. Recombinant DNA molecules are defined as either: *(a)* molecules that are constructed outside living cells by joining natural or synthetic DNA segments to DNA molecules that can replicate in a living cell; or *(b)* DNA molecules that result from the replication of those described in *(a).*

[27]Technically, the NIH guidelines only apply to institutions receiving monetary support from NIH. All federal agencies have required their own scientists to comply with the guidelines, however, and federal agencies other than NIH that fund rDNA research also require their grantees to comply with them.

[28]"Guidelines for research . . .", 46,272-73, sect. IV-D-1.

[29]In part VI of the NIH guidelines, a mechanism is created to encourage voluntary compliance by the private sector as well as a parallel system of administrative monitoring, modified to protect proprietary information. Appendix K, which concerns physical containment recommendations for large-scale uses of organisms containing rDNA, is also particularly relevant to the private sector.

[30]Commercial Biotechnology . . ..

The guidelines create four biosafety levels (BL) that relate to the degree of estimated biohazard and set forth the laboratory practices and techniques, safety equipment and laboratory facilities appropriate for the operations performed and the hazards posed by the agents. BL4 prescribes the most stringent containment conditions and BL1 the least stringent.[31]

The guidelines also provide for biological containment. This containment is based upon natural barriers that limit the infectivity of a vector to specific hosts or limit the dissemination and survival of the vector or the host in the environment.[31] Two levels of biological containment are defined.[32]

The guidelines create an administrative framework for monitoring that specifies the responsibilities of scientists, their institutions and the federal Government. The primary responsibility for insuring compliance lies with the institutions and the scientists who are doing the research. The institution must establish an institutional biosafety committee (IBC) meeting certain requirements, appoint a biological safety officer if certain experiments are done, ensure appropriate training and implement health surveillance, if appropriate. The principle investigator has the initial responsibility for determining and implementing containment levels and other safeguards and for training and supervising the staff.[33]

The IBC oversees all rDNA work at the institution, and, in fact, often acts as a general safety committee for the institution. It must consist of at least five members who collectively have the expertise to evaluate the risk of rDNA experiments. Two members must be otherwise unaffiliated with the institution and represent the community's interests pertaining to health and the environment. Institutions are encouraged to open IBC meetings to the public, and minutes of IBC meetings and certain other documents must be made available to the public on request. Institutions must register the IBC with NIH by providing information about its members.[34]

At the federal level, the responsible parties are the Director of NIH, the NIH Recombinant DNA Advisory Committee (RAC), the NIH Office of Recombinant DNA Activities (ORDA) and the Federal Interagency Advisory Committee on Recombinant DNA Research (Interagency Committee). The Director of NIH is the final decision-maker under the Guidelines. For major actions, the Director must seek the advice of RAC and must provide the public or other federal agencies with at least 30 days to comment on proposed actions. Every action taken by the Director must present "no significant risk to health or the environment".[35] RAC is a diverse group of experts that meets three or four times a year to advise the Director of NIH on major technical and policy issues. ORDA performs the administrative functions of NIH under the guidelines. Additional monitoring is provided by the Interagency Committee. Composed of representatives of approximately 20 agencies, this committee coordinates all federal rDNA activities, and its members are non-voting members of RAC.[36]

---

[31]"Guidelines for research . . .", 46,267, sect. II; and 46,279-85, appendix G.

[32]*Ibid.*, 46,286-88, appendix I.

[33]*Ibid.*, 46,269-70, sect. IV-B-1.

[34]*Ibid.*, 46,270, sect. IV-B-2.

[35]*Ibid.*, 46,271-73, sect. IV-C.

[36]The federal Government has proposed a major restructuring of its oversight of biotechnology. See, "Proposal for a coordinated framework for regulation of biotechnology", 49 Fed. Reg. 50,856-907 (1984).

The NIH guidelines classify all experiments into four categories (see annex I): those requiring RAC review and NIH approval before initiation (class III-A); those requiring IBC approval before initiation (class III-B); those requiring IBC notification at the time of initiation (class III-C) and those that are exempt (class III-D).[37] Containment levels are specified for each class except the one requiring NIH approval, where containment is set on a case-by-case basis. Class III-A covers experiments involving the formation of rDNA-containing genes for the synthesis of certain toxins that are lethal to vertebrates, the deliberate release of organisms containing rDNA into the environment, the transfer of drug resistance to certain micro-organisms under certain conditions, and the transfer of rDNA into human subjects. Class III-B covers experiments involving certain pathogenic organisms, whole animals or plants, or more than 10 litres of culture (except for certain exempt experiments). Class III-C is a catch-all for experiments that have not been placed in the other categories. The exempt category (class III-D) covers an estimated 80-90 per cent of all rDNA experiments.[38] Examples include most work with *E. coli* K-12, *S. cerevisiae* and asporogenic *B. subtilis* host-vector systems.

The United States Government is currently reconsidering its approach to the regulation of biotechnology. This process could result in major changes in the guidelines and a very different role for NIH. On 31 December 1984, the Office of Science and Technology Policy (OSTP) of the Executive Office of the President of the United States proposed a co-ordinated framework for regulation of biotechnology (hereinafter the "co-ordinated framework").[39] The co-ordinated framework envisions five agencies playing the major role in the regulation of biotechnology. NIH and the National Science Foundation (NSF) would oversee grant-supported research. The Environmental Protection Agency, the Food and Drug Administration and the Department of Agriculture would regulate the products of biotechnology. Each agency would maintain a scientific advisory board similar to RAC. There would also be a parent Biotechnology Science Board (BSB) composed of two representatives from each of the agency advisory boards and other members of the public. A separate inter-agency committee would co-ordinate decision-making and communication between the agencies and sort out matters of jurisdiction.

A number of criticisms were voiced against this two-tier mechanism. BSB in particular was seen as being redundant and cumbersome and unable to preserve confidentiality.

In response, OSTP developed a new arrangement, which it adopted on 14 November 1985. Instead of BSB, OSTP created an all-governmental body called the Biotechnology Science Coordinating Committee (BSCC) under the Federal Coordinating Counsel for Science and Technology (FCCST), which is housed in OSTP. BSCC will act as a clearing-house on scientific issues, evaluate agency review procedures, evaluate broad scientific issues, identify gaps in scientific knowledge and indirectly act as a forum for public concern.[40]

---

[37] "Guidelines for research . . .", 46,267-69, sect. III.

[38] *Commercial Biotechnology . . .*, p. 551.

[39] "Proposal for a coordinated framework . . .", 50,856 (1984).

[40] "Coordinated framework for regulation of biotechnology; establishment of the Biotechnology Science Coordinating Committee", 50 Fed. Reg. 47,174-95 (1985).

Japanese rDNA guidelines

Working involving rDNA in Japan is governed by two sets of virtually identical guidelines. One, promulgated by the Ministry of Education on the recommendation of the Science Council, governs academic research. The other, promulgated by the Science and Technology Agency, governs all other work with rDNA, including industrial work. Historically, the Japanese guidelines have been among the most restrictive in the world.

Each research institute is required to have laboratory supervisors, a safety committee and a safety officer. The head of each institution is also charged with specific duties in supervising the rDNA work. The laboratory supervisor must submit plans of experiments and changes in plans to the head of the research institute for approval. The head of the institution then consults with the safety committee to determine whether the plans comply with the guidelines, what training will be necessary and other issues relevant to the safety of the research. The safety officer's role is to monitor the safety of the ongoing work and to make appropriate reports to the safety committee.

The guidelines require physical and biological containment based upon the perceived risk of each experiment. The risk is assessed principally according to a phylogenetic scale, in which DNA donor organisms closer phylogenetically to humans are considered riskier. Risk is also assessed according to the biological characteristics of the source of the DNA, the purified or unpurified nature of the DNA, the size of the clone number and the scale of cultivation. The required physical and biological containment measures are similar to those of the NIH guidelines.[41]

The guidelines prohibit several types of experiments without prior government approval. These include the following: *(a)* the use of unapproved host-vector systems; *(b)* the use of certain procaryotic or eucaryotic donors; *(c)* the cloning of genes coding for toxins lethal to vertebrates; *(d)* certain large-scale work; and *(e)* the intentional release of rDNA-containing organisms into the environment.

Special guidelines for rDNA experiments using plants or animals were approved in February 1984. The plant guidelines require the physical isolation of rDNA-containing plants, including appropriate measures for preventing pollen or seed dispersal, and prior government approval for the transfer of the plants to other institutions. The animal guidelines require the isolation of the experimental animals and proper disposal of experimental materials. They prohibit the introduction of rDNA into human or primate eggs and the breeding of genetically engineered animals.

Recombinant DNA guidelines of the United Kingdom

The United Kingdom guidelines for rDNA research are similar to the NIH guidelines in broad conceptual terms, but they differ with respect to scope, risk assessment and enforcement. Also, the United Kingdom guidelines are actually a collection of many different advisory notes on different topics, whereas the NIH guidelines are a single document.

The advisory notes were promulgated by the Genetic Manipulation Advisory Group (GMAG) under the authority of the Health and Safety at

---

[41]*Commercial Biotechnology . . . ,* pp. 552-555.

Work etc. Act 1974[42] and the Health and Safety (Genetic Manipulation) Regulations 1978.[43] They apply to all rDNA research in the United Kingdom. Although they are not regulations, they have an implied legal authority because, under the Act, an employer can be prosecuted for failure to use safe practices in the work-place. These guidelines embody what are considered to be safe practices with respect to rDNA research.

The genetic manipulation regulations, on the other hand, are binding upon all persons carrying out rDNA work. They provide that such persons cannot do such work until they have notified the Health and Safety Executive and GMAG.[44]

GMAG had been located in the Department of Education and Science. In early 1984, GMAG was disbanded and responsibility for monitoring genetic manipulation was transferred to the newly formed Advisory Committee on Genetic Manipulation (ACGM) in the Health and Safety Executive (HSE), which is responsible for worker health and safety. This advisory committee is one of several advisory committees in HSE, all of which are composed of representatives from industry, labour and the public in equal numbers. All but one of the GMAG advisory notes remain in effect, and ACGM has promulgated additional advisory notes. Its main focus is now on industrial and environmental uses of genetically manipulated organisms. This restructuring of authority over genetically engineered organisms represents a decision that such work does not present particularly unique risks that should be dealt with outside of the traditional agency for dealing with any potential risks presented in the workplace, i.e. HSE.[45]

The United Kingdom guidelines categorize rDNA experiments on the basis of assigned risk values. The risk values are determined by considering three factors: *(a)* access; *(b)* expression; and *(c)* damage.[46] The guidelines establish four progressively more restrictive physical containment levels based upon the perceived risk of experiments. Facilities for the highest two levels must be examined by HSE inspectors before any rDNA research can be conducted to ensure that the requirements are met. The guidelines also adopt the two-level biological containment approach of the United States and most other countries, which is based upon the degree of disability of the host-vector system being used.

Special rules have been developed for rDNA research that involve the introduction of foreign nucleic acids into higher plants or into any plant pest. Laboratory and greenhouse containment requirements are specified, and

---

[42]Health and Safety at Work etc. Act, 1974, *Statutory Instruments,* chap. 37.

[43]United Kingdom of Great Britain and Northern Ireland, SI 1978 No. 752 Her Majesty's Stationery Office.

[44]Once the initial notification has been made, notification for the category of experiments subject to the least restrictions need only be retrospective. Genetic Manipulation Advisory Group, *Third Report of the Genetic Manipulation Advisory Group* (London, 1982), p. 8.

[45]Telephone interview with M. G. Norton, First Secretary (Science), British Embassy, Washington, D.C., 30 October 1985.

[46]*Commercial Biotechnology . . . ,* p. 553. "Access" is the possibility that escaped organisms will enter the human body and reach susceptible cells. "Expression" is the possibility that a foreign gene incorporated into the gene sequence of an organism will be able to carry on or "express" its normal function, such as the secretion of a toxin that the organism formerly did not secrete. "Damage" is the chance that a new gene sequence will cause physiological damage in the body to which it gains access once it is expressed.

experiments involving the genetic manipulation of plant pests require a license from the Ministry of Agriculture, Fisheries and Food.[47]

As in the United States, the administrative framework for implementing the United Kingdom guidelines relies on institutional and governmental supervision. ACGM and HSE must be notified of certain experiments before they are undertaken.[48] In addition, each institution conducting rDNA research is required to have certain personnel responsible for reviewing the research, forwarding notifications to ACGM and suggesting health and safety actions that the institution might take.[49]

European Economic Community rDNA guidelines

The European Economic Community (EEC) issued guidelines for rDNA research in June 1982 in the form of a non-binding recommendation by the Council of the European Communities to member States.[50] It is suggested in the guidelines that any laboratory wishing to conduct rDNA research, except for research of a very low risk potential,[51] should notify the competent national or regional authority in the member State. The notification would include information about the experimental protocol, the protective measures to be taken and the general education and training of the staff working on the experiment or monitoring it. Such notification was thought to be desirable because it would create records that would be helpful in the highly unlikely event of an accident. It is also recommended in the guidelines that the authority receiving the notification should protect the confidentiality of the information submitted.

General laboratory biosafety guidelines

As previously mentioned, a consensus seems to have emerged among experts that rDNA and other genetic engineering techniques present no special risks in themselves and, therefore, ought to be governed by standard good laboratory practices. Such practices are generally learned by young scientists in the course of their training from more senior scientists, but there are also written guidelines that specify these practices in microbiological and biomedical laboratories. Two examples of such guidelines are the WHO *Laboratory Biosafety Manual*[52] and a publication developed jointly by the United States Centers for Disease Control and National Institutes of Health entitled *Biosafety in Microbiological and Biomedical Laboratories* (hereinafter cited as CDC/NIH *Biosafety Manual*).[53]

---

[47]Genetic Manipulation Advisory Group, *Genetic Manipulation of Plants and Plant Pests*, GMAG Note No. 13 (London, 1980).

[48]The advice of HSE and ACGM must be obtained for all work, including all large-scale work except that in category I (lowest risk experiments), before the work commences. Genetic Manipulation Advisory Group, *Revised Guidelines for the Categorization of Recombinant DNA Experiments*, GMAG Note No. 14 (London, 1981).

[49]*Commercial Biotechnology . . .*, p. 553.

[50]Council of the European Communities, "Council recommendation of 30 June 1982, No. 82/472/EEC, concerning the registration of work involving deoxyribonucleic acid (DNA)", *Official Journal of the European Communities* (L213), vol. 25, 21 July 1982, pp. 15-18.

[51]The term "very low risk potential" is not defined in guidelines, but it is indicated that this is to be determined by the competent national authorities.

[52]*Laboratory Biosafety Manual . . .*.

[53]*Biosafety in Microbiological and Biomedical Laboratories . . .*.

The WHO *Manual* is a comprehensive document on all aspects of laboratory biosafety. It is organized into three major parts: guidelines; laboratory practice and management; and a guide to biosafety equipment. In addition, there is an extensive bibliography of national codes and other relevant materials with respect to various aspects of laboratory biosafety.

The *Manual* is a synthesis of advice formulated by a number of international working groups of experts established under the WHO Special Programme on Safety Measures in Microbiology. It recognizes three main factors that can effect its international application. First, risks ascribable to certain biological agents vary in different countries; what may be an important pathogen in one part of the world may be a lesser one in another. Secondly, the varying levels of development in laboratory facilities throughout the world are such that it is essential for any proposed safety measures to fit the available resources. Thirdly, the needs of laboratory personnel in different countries vary according to their training and the work they are required to do. Thus, recommended procedures must be adaptable to a wide variety of educational backgrounds and laboratory practices. Accordingly, the *Manual* can be a source document from which laboratory manuals applicable to local circumstances can be derived.[54]

Certain general principles underlie the guidelines and practices set forth in the *Manual.* First, infectious organisms can be classified according to the risk they present to individuals in the laboratory and to the community at large. Secondly, the risk can be classified in various levels from low to high. The guidelines and practices are geared to these increasing levels of risk. Thirdly, the containment of the organisms is the principal means of addressing the risks. Fourthly, sound microbiological practices must be inculcated in the scientists, technicians and other support staff.

The manual covers four risk groups. Group I contains micro-organisms with a low risk to individuals and the community. Group II contains organisms with a moderate risk to individuals but a limited risk to the community. Group III contains organisms with a high risk to individuals but a low risk to the community. Group IV contains micro-organisms with a high risk to both individuals and the community.

The guidelines section of the *Manual* is actually a collection of four different groups of guidelines for four different types of laboratories: the basic laboratory; the containment laboratory; the maximum containment laboratory; and the genetic engineering laboratory. They cover such topics as practices, design of facilities, equipment, medical surveillance, training, procedures for handling materials, emergency procedures, decontamination and disposal of materials, and animal facilities. The guidelines for the genetic engineering laboratory occupy only one page of the *Manual* because, as mentioned previously, the experts who prepared the *Manual* have concluded that the risks presented are no greater than those associated with known pathogens and, therefore, do not necessitate special laboratory design or practice.[55] The *Manual* does, however, contain a table of proposed safety levels for work with rDNA to aid in the selection of suitable standard laboratory facilities and practices as outlined in other sections of the guidelines.

---

[54]Interview with Vinson R. Oviatt, Director, WHO Programme on Safety Issues in Biotechnology, Geneva, Switzerland, 9 September 1985.

[55]*Laboratory Biosafety Manual . . .*, p. 30.

The section of the *Manual* entitled "Laboratory practices and management" specifies various practices for use in particular situations and general monitoring procedures for the laboratory. It details appropriate techniques for scientists, technicians and other support staff, such as custodians. In a section on training, it outlines course topics, structure and materials. Emergency procedures are specified for various situations.

The implementation of all these practices are accomplished by an appropriate safety management team. It is suggested that a biosafety officer should be appointed whenever possible to ensure that safety policies and programmes are followed throughout the laboratory. It is also suggested, at least for larger institutions, that a biosafety committee should be set up to recommend a safety policy and programme and to formulate or adopt a code of practice or a safety manual. The composition and duties of these various parties are discussed.

Finally, the *Manual* contains an extensive guide to safety equipment. The appropriate equipment for appropriate levels of hazards is identified, and the equipment and how equipment itself can create a hazard are discussed.

The CDC/NIH *Biosafety Manual,* which covers standard and special microbiological safety practices, safety equipment and facilities, is similar to the WHO *Manual.* It too is built on the concept of categorizing infectious agents and laboratory facilities into four levels and on the concept of containment of infectious agents. Primary containment is based upon techniques and safety equipment, and secondary containment is based upon laboratory design and operational practices.

Four biosafety levels are defined, which consist of combinations of laboratory practices and techniques, safety equipment and laboratory facilities appropriate for the operations performed and the hazards posed by the infectious agents. Biosafety level 1 concerns work done with defined and characterized strains of viable micro-organisms not known to cause disease in healthy adult humans. Biosafety level 2 covers work with indigenous, moderate risk agents present in the community and associated with human disease of varying severity. Biosafety level 3 covers work done with indigenous or exotic agents where there is a potential for infection by aerosols and the disease may have serious or lethal consequences. Biosafety level 4 covers work with dangerous and exotic agents that pose a high individual risk of life-threatening disease.[56] For each of the biosafety levels, standard microbiological practices, special practices, containment equipment and appropriate laboratory facilities are defined.

Vertebrate animal biosafety levels are also defined and discussed. As with the microbiological safety levels, four animal biosafety levels are defined and appropriate practices, equipment and facilities are discussed with respect to each.

Finally, particular biosafety levels for particular infectious agents and infected animals are recommended. The CDC/NIH *Biosafety Manual* does not specifically cover genetic engineering techniques.

---

[56] *Biosafety in Microbiological and Biomedical Laboratories . . .,* pp. 6-7.

16

Evaluation of current regulations

Several common principles and themes run through the various guidelines dealing with rDNA or general microbiological techniques. A scientific assessment of the risk presented by the particular organisms involved is the starting point. Levels of risk are defined, and the organisms are classified within those levels. Appropriate practices, equipment and facilities are determined for particular levels of risk so that the physical containment of organisms increases as the degree of risk increases. In some guidelines, such as the NIH rDNA guidelines, the concept of biological containment is also employed. The guidelines also are grounded upon the principle that good microbiological practices must be an integral part of the general training of all laboratory personnel.

The level of administrative supervision also increases with increasing risk. Government authorities must be notified before initiation of experiments for work in the higher-risk categories. Biosafety committees are recommended for institutions sponsoring rDNA work, and biosafety officers are usually recommended, too.

Another aspect of the guidelines is their flexibility. They can be adapted to new circumstances or new data. If experience shows that the risk of a particular organism was less than originally perceived, that organism can be placed in a lower-risk level. On the other hand, if special circumstances warrant additional precautions when working with a particular organism, those precautions can be taken. The flexibility of the guidelines also allows them to be adapted to the different needs of different countries.

The current monitoring mechanism of voluntary, self-regulation in the form of guidelines appears to be adequate for dealing with the risks presented to laboratory workers by micro-organisms, whether genetically engineered or not. The guidelines for good laboratory practices in the microbiological laboratory have been developed over and are based upon several decades of experience. Even the newer guidelines that are focused solely on rDNA are the result of over 10 years of experience with that technique in the laboratory. During this time, there have been no reports of illnesses or injuries attributed to the rDNA technique. Most experts believe that laboratory work with rDNA presents no risks beyond those already inherent in the biological materials and systems being used. Some experts, however, have expressed concern about the possibility of activating otherwise unexpressed genes that make harmful products. Given the current views of the experts on the risks presented by rDNA and the history of increasing relaxation of the restrictions in the original rDNA guidelines, it appears likely that guidelines directed specifically toward rDNA or other types of genetic engineering will eventually be subsumed within the framework of more general guidelines directed towards good practices in the biological laboratory. Thus, any guidelines that ICGEB might develop should be from the perspective of appropriate practices for the biological laboratory with sections covering genetic engineering techniques, as appropriate.

II. Risks and Regulation of large-scale operations

Potential risks

The question of the nature and extent of risks presented by large-scale operations involving genetically engineered organisms, i.e. large fermenters with contained organisms, appears to be less settled than in the case of laboratory-scale operations. This state of affairs is probably due to the fact that biotechnology companies have had much less experience with large-scale uses of genetically engineered organisms, although this situation is quickly changing. To date, there have been no reports of harm to workers resulting from actual or potential exposure to genetically engineered organisms or their products. Similarly, there have been no reports of illness or injury to the communities surrounding biotechnology facilities or of adverse impacts on the environment. Nevertheless, there is concern about the possible risks presented by such operations.

It is clear that large-scale fermentation operations will produce large quantities of biowastes. These wastes can include water, reagents and micro-organisms. They will have to be disposed of safely and in such a way as to limit adverse environmental impacts. There is substantial technical experience and legal authority for dealing with biowastes. Nevertheless, there are some special issues that should be considered with respect to genetic engineering biowastes from large-scale operations.

Some people have argued that large-scale operations may be more risky than laboratory-scale operations. The larger amount of material involved might increase the probability of exposure of workers, especially in the case of a spill. In addition, factory workers may be less well-informed of the hazards of particular organisms and may be less well-trained in safety procedures than laboratory personnel.

On the other hand, other people have argued that large-scale facilities will be safer than laboratory facilities. One reason is their inherently better containment features. Instead of glass, large-scale facilities are constructed from metal, plastic and other sturdy materials. In addition, when a high degree of product purity is desired, the facilities and procedures for attaining that high purity necessarily involve a high degree of containment of organisms and products. As a standard practice, containment vessels are sterilized at the beginning of the process. At the end of the process, the micro-organisms are usually destroyed to facilitate the removal and purification of the product. Another reason is the fact that the pharmaceutical industry has had many decades of experience with micro-organisms, including pathogens, and has developed appropriate techniques for safely handling them on a large scale.

An early attempt to consider the possible risks presented by large-scale applications of genetic engineering and to suggest ways of dealing with any risks was made by an *ad hoc* working group of individuals from the Centers for Disease Control (CDC) and the National Institute for Occupational Safety and Health (NIOSH).[57] In its report, the working group identified three types of hazards: microbial hazards; product hazards; and reagent hazards.

As to the first hazard, the group noted the existence of probabilistic arguments that workers may be colonized or infected by modified organisms. It noted that physical and biological containment would be the defence against such colonization or infection and concluded that the health hazards of exposure to altered micro-organisms appeared to be minimal because of the current use of highly attenuated microbial species.

The group found a greater degree of hazard from exposure to biologically active products of the genetically engineered micro-organisms. In support of this conclusion, it noted instances where exposure to products in other sectors of the pharmaceutical industry had produced a spectrum of illnesses. It also stated that workers would be at risk for sensitization to microbial proteins and peptides generated by fermentation and extraction, citing the high frequency of sensitization to protein enzymes among workers engaged in the commercial production of enzyme detergents. Asthma was cited as the most serious health consequence of such sensitization, but the group stated that dermatitis and allergic rhinitis might also be expected to occur.

As to the third type of hazard, it was noted that solvents and other chemical reagents would be used extensively in the extraction, separation and purification of the products produced by large-scale fermentation. The hazards of some of these solvents were well-known and standard procedures would have to be adopted for addressing them.

The working group concluded that the medical surveillance of biotechnology workers would be prudent medical practice in view of the lack of information concerning the nature and severity of health hazards that might be associated with industrial applications of the new biotechnology. Such surveillance should include: *(a)* a pre-employment examination with the collection of baseline serum; *(b)* periodic follow-up; *(c)* the evaluation of all illnesses causing 48 hours absence from work; *(d)* epidemiologic studies; *(e)* the periodic evaluation of data; and *(f)* regular communication of results to management and workers.

The report has been criticized, and its findings have not been implemented by the Occupational Safety and Health Administration (OSHA), the United States agency that is primarily responsible for worker health and safety. In particular, the need for medical surveillance and the ability to construct a meaningful programme have been questioned when it comes to searching for conjectural risks.[58]

In a more recent report, the European Federation of Biotechnology (EFB) Working Party on Safety in Biotechnology came to a more positive conclusion with respect to the risks presented by biotechnology and the means for

---

[57]P. J. Landrigan and others, "Medical surveillance of biotechnology workers: Report of the CDC/NIOSH *Ad Hoc* Working Group on Medical Surveillance for Industrial Applications of Biotechnology", *Recombinant DNA Technology Bulletin,* No. 133, 1982.

[58]*Commercial Biotechnology . . .,* p. 373.

addressing those risks.[59] In its report, the Working Party concluded that biotechnology is safe if properly practised.[60]

Certain problems specific to biotechnology are defined in the report: the pathogenicity of some micro-organisms; the problems associated with biologically active microbial products; the problems associated with handling bulk quantities of micro-organisms; and the stability and purity of process strains. For the first problem area, it is concluded that when pathogenic micro-organisms are used, they are being contained to such an extent that experience had shown that risks to humans, animals and plants are minimal.[61] The authors also proposed a classification of micro-organisms according to pathogenicity. With respect to the problems associated with biologically active microbial products, the authors concluded that the quality control systems in the biotechnology industries were sufficient to contain or remove the products in question. With respect to bulk quantities of micro-organisms, the authors simply noted that such micro-organisms could remain after the desired product had been separated but that those were handled by fairly standard means. Stability and purity problems were really directed towards the quality of the product rather than worker health and safety. In the EFB report, it was noted that such problems would be identified quickly by a manufacturer in the interest of maintaining an efficient and profitable process.

Recommendations for future action are also made in the EFB report. One is that efforts should be directed towards reducing the number of processes that employ pathogenic micro-organisms, such as by transferring genetic information related to process needs from harmful to harmless organisms. In cases where the use of hazardous organisms was unavoidable, it was recommended, among other things, that improved containment techniques should be developed. It is also recommended that regulations and guidelines should be harmonized.[62]

A WHO working group on the health impacts of biotechnology considered the issue of worker health and safety, among other issues.[63] It concluded that "at present, biotechnology appears to possess no risks that are fundamentally different from those faced by workers in other processing industries".[64] The working group went on to recommend that occupational exposure to micro-organisms or parts of micro-organism should be monitored for indications of allergic reactions and hypersensitivity and that the control of the working environment should be carried out in line with equivalent existing industries. It further advised that appropriate medical surveillance of all workers in biotechnology industries should be conducted. Finally, it recommended that steps should be taken to collect data from laboratories and production units to monitor the health and safety of workers and the influence on the environment.[64]

---

[59]M. Kuenzi and others, "Safe biotechnology: general considerations", *Microbiol. Biotechnol.* 21 Appl., No. 1 (1985).

[60]In the report, biotechnology is defined as the integrated use of biochemistry, microbiology and engineering sciences in order to achieve the technological (industrial) application of the capabilities of micro-organisms, culture tissue cells and parts thereof. As such, the definition includes the old as well as the new biotechnology.

[61]Kuenzi, *op. cit.,* p. 4.

[62]*Ibid.,* pp. 4-5.

[63]*Health Impact of Biotechnology . . ..*

[64]*Ibid.,* p. 26.

The WHO working group also identified the disposal of biological wastes as a topic that needed attention. It noted that an increase in the use of biological processes was unlikely to pose problems of waste treatment beyond those inherent in modern methods of sewage disposal and waste-water treatment in the chemical industry. It noted that potential health problems might arise from aerosols and dusts in addition to aqueous effluents and semi-solid sludge effluents from treatment plants. It also noted the possibility of thermo-pollution resulting from the discharge of heated effluents. For the small number of processes that were likely to involve known pathogens, it recommended that appropriate containment facilities should be required and waste should be sterilized. The group also stated that appropriate monitoring procedures necessary to detect accidental leakage or discharge from such processes should be implemented, and it noted the need for emergency procedures. Because of the considerable volume of water involved in biotechnological processes, it was important for treatment to be carried out in a manner that minimized the environmental impact of the discharge. For that reason, water reuse was encouraged. Health officers at the WHO Regional Office for Europe continue to see the waste disposal issue as an important one that needs much attention with respect to the biotechnology industries.[65]

Despite its general conclusion about the safety of biotechnological processes for workers, the WHO working group noted that it was appropriate to continue to assess whether biotechnological techniques presented potential health problems and to monitor such techniques and processes for long-term adverse effects.[66]

Risk assessment

Risk assessment of the large-scale applications of biotechnology is at a very early stage of its development, but the basic tools for such risk assessment exist. Several publications concerning risk assessment for other technologies are cited in the WHO report. The working group felt that the methods reported in those publications could be adapted for biotechnology and that some of the methods of the chemical industry for identifying and controlling chemical hazards could be applied to biotechnology.[67]

There has been little experience in applying these tools to large-scale biotechnological processes, however. According to Jorma O. Jarvisalo, Regional Officer for Occupational Health, WHO Regional Office for Europe, there is not yet sufficient knowledge to attempt risk assessment in the area of occupational health issues, including monitoring workers in biotechnology plants. In other words, no one knows what long-term impacts to look for. He did believe, however, that workers could be monitored for allergic reactions to particular products of biotechnology. He further stated that basic occupational health and safety principles and engineering techniques could be applied to controlling and monitoring plants and equipment.[68] Lech J. Piekarski of UNEP

---

[65]Interview with Ian Waddington, Michael Suess and Jorma O. Jarvisalo, Health Officers, WHO Regional Office for Europe, Copenhagen, Denmark, 6 September 1985.

[66]*Health Impact of Biotechnology . . .*, p. 8.

[67]*Ibid.*, p. 28.

[68]Interview with Jorma O. Jarvisalo, Regional Officer for Occupational Health, WHO Regional Office for Europe, Copenhagen, Denmark, 6 September 1985.

agreed. He noted that the first step in risk assessment is hazard identification and suggested that there might be sufficient data for hazard identification to be carried out.[69]

Regulation of large-scale genetic engineering processes

As with genetic engineering in the laboratory, there are no statutes or regulations specifically directed towards large-scale processes involving genetically engineered organisms. National guidelines on rDNA techniques, however, are quasi-regulatory in nature and cover large-scale processes to varying degrees. In addition, there are various guidelines for general good manufacturing processes. Finally, most countries have worker health and safety laws and pollution control laws. In Japan, the United States and the western European countries, these laws provide broad protection for workers and the environment and encompass genetic engineering activities. They also provide the authority for Governments to address any particular risks raised by genetic engineering by promulgating special regulations.

Japanese guidelines

Until late 1983, large-scale (more than 20 litres of culture) work with rDNA-containing organisms in Japan had been effectively prohibited. Special permission had to be granted by the Ministry of Education, which granted it only rarely. This restriction was removed for most work, and a two-tier containment scheme was devised. LS_1 and LS_2 containment levels are specified for work similar to the two lowest levels of physical containment (P-1 and P-2) for comparable small-scale work. LS_1 facilities are similar to those for conventional micro-organism laboratories. LS_2 facilities are covered by more restrictive rules, such as those for facilities handling aetiologic agents. Large-scale experiments with rDNA-containing organisms that would require higher physical containment at the laboratory-scale (P-3 or P-4) still require government approval before initiation.

The Ministry of Trade and Industry has been developing special comprehensive guidelines for the industrial application of rDNA technology.

Guidelines of the United Kingdom

Under the United Kingdom guidelines, the large-scale use of the products of genetic manipulation is defined as work involving volumes of 10 litres or more and is subject to special rules. HSE and ACGM review proposals to conduct such work on a case-by-case basis. The underlying rationale is that scale-up involves a significant change from the manipulation of genetic material to the use of the resulting genetically engineered organism. Therefore, any risks involved will not necessarily remain the same, and a different type of assessment will be required for these changed circumstances. It is noted in the guidelines that vaccine and antibiotic production are two well-established

---

[69]Interview with Lech J. Piekarski, United Nations Environment Programme, Nairobi, Kenya, 13 September 1985.

industrial biological processes that provide experience for assessing the large-scale use of genetically engineered organisms. It is further noted that there is little to distinguish these processes from those of chemical engineering in which effective design, efficient fabrication and skilled operation provide a very high degree of containment.[70]

HSE has recognized the commercial importance of genetic engineering by establishing special confidentiality requirements for work that raises questions about commercial property or patents. Although the confidentiality arrangements may vary from case to case, HSE and ACGM generally treat any materials so labelled as confidential. Members of ACGM who have commercial interests in rDNA work are prohibited from seeing such material or taking part in discussions about it.[71]

Although it is noted in the United Kingdom guidelines that there are no known health hazards specific to genetic manipulation, fairly stringent health monitoring of workers is required in the guidelines. The reason is the existence of conjectural risks and the fact that, as with microbiological work that does not involve genetic manipulation, the micro-organisms that are used may be capable of infecting humans. A system of health monitoring is suggested involving: health cards for each worker; an initial medical examination; the collection and storage of serum samples; the maintenance of appropriate records; and a follow-up on unexplained illnesses. For workers in facilities classified as containment categories III or IV, annual health reviews are recommended. A further requirement is the appointment of a supervisory medical officer for each laboratory who should be experienced in public health, infectious diseases or occupational medicine.[72]

NIH guidelines

In the NIH guidelines, work involving more than 10 litres of culture is considered to be a large-scale use of organisms containing rDNA.[73] For such work, the IBC are permitted to determine the appropriate containment but appendix K to the guidelines on the physical containment for large-scale uses of organisms containing rDNA molecules should be used where appropriate. Three physical containment levels for large-scale research or the production of viable organisms containing rDNA molecules are set out in appendix K, along with a discussion on how the appropriate physical containment level is selected and what engineering and processing requirements must be met. It is also suggested that the institution should appoint a biological safety officer, whose duties are specified in section IV-B-4 of the guidelines. It is also suggested that the institution should establish a health surveillance programme for work involving organisms that require BL3 containment at the laboratory scale. Such a programme should include: pre-assignment and periodic physical and medical

---

[70]Genetic Manipulation Advisory Group, *Large Scale Uses of the Products of Genetic Manipulation–Work Involving Volumes of Ten Litres or More,* GMAG note No. 12 (London, 1979).

[71]*Ibid.*; Genetic Manipulation Advisory Group, *Information and Advice on the Completion of Proposal Forms for Centers (Part A) and Projects (Part B),* GMAG revised note No. 7 (London, 1979).

[72]Genetic Manipulation Advisory Group, *Health Monitoring,* GMAG note No. 6 (London, 1980); and *Revised Guidelines for the Categorization of Recombinant DNA Experiments,* GMAG note No. 14 (London, 1981).

[73]"Guidelines for research . . .", 46,266-69, sect. III-B-5.

23

examinations; the collection, maintenance and analysis of serum specimens for monitoring serologic changes that may result from employees' work experience; and provisions for the investigation of any serious, unusual or extended illness of any employees to determine possible occupational origin.

Since 1980, NIH has had an *ad hoc* group known as the Large Scale Review Working Group to advise RAC on procedures and facilities pertaining to large-scale operations. The group is composed of certain members of RAC, members from other federal agencies and other experts as appropriate. The group has periodic meetings to discuss issues pertaining to large-scale containment and makes recommendations to RAC as appropriate.

General good manufacturing practice guidelines

The pharmaceutical industry has many years of experience in dealing with large-scale cultures of micro-organisms, including pathogenic ones. It would appear that these practices are directly applicable to genetically engineered organisms, although some modifications may be necessary. There are at least three written sources of good manufacturing practices (GMPs) that could be applied to large-scale fermentation processes involving genetically engineered organisms, even though they are directed specifically towards the production of pharmaceuticals. Two are WHO publications,[74] and another is a series of regulations of the United States Food and Drug Administration (FDA).[75] The latter is attached as annex II.

One of the WHO publications covers biological substances, such as vaccines, and the other covers drugs. Both have general requirements for manufacturing establishments, covering such topics as equipment, personnel, operations and quality control. WHO is currently in the process of using these documents to develop good manufacturing practices for biotechnology. Significant work will have to be done in developing suggested practices for worker health and safety, since the current GMPs are more oriented towards quality control.[76]

The GMPs of FDA are comprehensive regulations governing the manufacturing, processing, packaging and holding of drugs. They cover the design of buildings and facilities, equipment design, operation and maintenance, production and process controls, and the responsibility of personnel.

Statutes governing worker health and safety

Most countries have some type of statutory protection for the health and safety of their workers. Perhaps the most comprehensive statutes are those found in some of the developed countries, in particular Japan, the United Kingdom and the United States. Each of these countries imposes general duties on employers to maintain safe work-places and to eliminate or control hazards.

[74]*Requirements for Biological Substances: Report of a WHO Expert Group,* Technical Report Series No. 23 (Geneva, World Health Organization, 1966); and *Quality Control of Drugs* (Geneva, World Health Organization, 1977).

[75]United States of America, 21 C.F.R. Parts 210 and 211 (1985).

[76]Interview with Vinson R. Oviatt, Director, Programme on Safety Measures in Microbiology, WHO, Geneva, Switzerland, 9 September 1985.

They provide the Government with the authority to promulgate regulations specifically directed towards genetic engineering product technology. Such regulations are likely to be primarily process- rather than product-oriented.

Japan

The Industrial Safety and Health Law of 1972[77] is the basic law of governing worker health and safety in Japan. It imposes health and safety obligations on employers that are comprehensive in scope but very general in actual language. Among these obligations is the duty to take necessary measures to prevent health impairments caused by substances, agents and conditions found in the work-place. The law vests broad discretion in the Japanese Ministry of Labor to determine when regulation is appropriate and what kind of precautions an employer must take. Employers who manufacture, import or use chemical substances may be subject to special requirements. All employees must undergo medical examinations, and the employers may also be required to provide special tests for employees engaged in harmful work. At the present time, there are no regulations specifically covering genetic engineering.[78]

The law includes a stringent enforcement mechanism. Substantial criminal penalties and fines are imposed for violations. For the most serious violations, offending employers may also be ordered to alter or close their operations.[78]

United Kingdom

The principal statute governing worker health and safety in the United Kingdom is the Health and Safety at Work etc. Act 1974.[79] It places certain obligations on virtually all employers and manufacturers. In general, employers must ensure as far as reasonably practicable that employees are not exposed to health and safety risks and must inform employees of any risks.

Regulations under the act are promulgated by the Secretary of State on the advice of the Health and Safety Commission. The Health and Safety Commission also supervises efforts to improve worker safety and health, makes necessary investigations and may approve codes of practice for particular industries.

Codes of practice are quasi-regulatory. The violation of a code is not a violation of the act *per se*, but it is evidence of a violation. There are no codes of practice for genetic engineering or biotechnology other than the GMAG guidelines for rDNA research.

HSE and local authorities enforce the act through appointed inspectors, who may issue notices that prohibit certain activities or that require remedial actions. Violators of the act are subject to civil and criminal penalties.

United States

In the United States, the agency primarily responsible for worker health and safety is OSHA, which is part of the United States Department of Labor. The authority of OSHA derives from the Occupational Safety and Health Act

---

[77]Industrial Safety and Health Laws, Law No. 57 of 8 June 1972, as amended by Law No. 28 of 1 May 1975 (Working Environment Measurement Law) and Law No. 76 of 1 July 1977; translation available in: Japan, Ministry of Labor, *Labor Laws of Japan* (Tokyo, 1980).

[78]*Commercial Biotechnology . . .*, p. 561.

[79]Health and Safety at Work etc. Act 1974, *Statutory Instruments*, chap. 37.

of 1970,[80] which creates a broad mechanism for protecting workers from hazards in the work-place. Section 5(a)(1) of the act requires employers to furnish their employees with a work-place "free from recognized hazards that are causing or are likely to cause death or serious physical harm". Section 5(a)(2) requires employers to comply with the safety and health standards set by the Secretary of Labor. Under a 1980 Supreme Court decision, the Secretary of Labor can promulgate permanent standards for toxic substances or harmful physical agents only after finding that the standard is "reasonably necessary and appropriate to remedy a significant risk of material health impairment".[81] Section 6(c) of the act permits the Secretary of Labor to promulgate emergency temporary standards after finding that employees are "exposed to great danger". Other sections grant OSHA the authority to require record keeping and medical surveillance and to enforce the act and its regulations through civil and criminal penalties. The statute also created the National Institute for Occupational Safety and Health to gather data, assess risks and recommend safety and health standards to OSHA.

In a recent policy statement, OSHA took the position that the employer's general duty to maintain a safe work-place as specified in section (5)(1) of the act, together with several specific standards promulgated by OSHA, provides an adequate and enforceable basis for the protection of the health and safety of employees in the field of biotechnology.[82] The Agency further stated:

"No additional regulation of work places using biotechnology appears to be needed at this time, since no hazard or hazards from biotechnology *per se* had been identified. However, if any of the new biotechnology processes cause hazardous working conditions that result in a significant risk of death or serious harm to workers, OSHA will consider regulating unless the worker exposure is effectively controlled under current OSHA standards or another agency has exercised its authority over health and safety matters for those working conditions."[82]

Elaborating upon its conclusion, the Agency noted that biotechnology processes, whether present in laboratories, pilot plants or industrial plants, usually involved conventional chemicals and processes that are already covered by OSHA regulations. Thus, the Agency stated that:

"The potentially hazardous character of some aspects of biotechnology is primarily from the chemicals used and not the biotechnology products. Therefore, the regulations that effectively regulate chemical exposures will usually ensure that biohazards, too, will be controlled."[83]

In the policy statement, the Agency also noted that employers were required to comply with occupational safety and health standards promulgated under the act. Standards identified as potentially applicable included: *(a)* specific ones dealing with air contaminants; *(b)* access to employee exposure and medical records; *(c)* hazard communication; *(d)* exposure to toxic chemicals in laboratories (currently under development); *(e)* respiratory protection; and *(f)* safety standards of a general nature, such as those dealing with working area, fire protection, electrical safety and material handling and storage.

---

[80]United States of America, 29 U.S.C. 651-678.

[81]United Sates of America, *Industrial Union Dept., AFL-CIO v. American Petroleum Inst.,* 448 U.S. 607, 639 (1980).

[82]United States of America, 50 Fed. Reg. 14,468 (12 April 1985).

[83]*Ibid.,* 14,469.

The policy statement also contains some recommendations with respect to biotechnological work. These include informing and instructing all personnel with respect to real and potential hazards, the formulation of emergency plans and immunization against known pathogens, as appropriate. Adequate training is also suggested.

The European Economic Community

The powers of EEC in the area of worker health and safety regulation are limited and indirect. It has attempted, however, to ensure at least minimal protection for most industrial workers. In 1980, EEC adopted a directive that required each member State to adopt a variety of measures to protect workers' health and safety when they were exposed to chemical, physical or biological agents likely to be harmful to health.[84] The required measures include the following:

(a) Limitations on the use of chemical, physical, or biological agents in the work-place;

(b) Limitations on the number of workers exposed or likely to be exposed to such agents;

(c) Engineering controls;

(d) Establishment of exposure limit values for such agents and methods of assessing their level;

(e) Safe working procedures and methods;

(f) Collective protection measures;

(g) Individual protection measures when exposure cannot reasonably be avoided by other means;

(h) Hygiene measures;

(i) Information for workers on potential risks associated with exposure to such agents, preventive measures workers should take, and precautions to be taken by the employer and the workers;

(j) Use of warning and safety signs;

(k) Surveillance of workers' health;

(l) Maintenance of current records of exposure levels, workers exposed and medical records;

(m) Emergency procedures;

(n) If necessary, general or limited bans on an agent from which protection cannot be adequately ensured.

The directive does not refer explicitly to genetic engineering techniques. Thus, the implementation of the directive and national worker health and safety laws with respect to genetic engineering processes is left to the discretion of each member State.

_____

[84]Council of the European Communities, "Council directives of 27 November 1980, No. 80/1107/EEC, on the protection of workers from the risks related to exposure to chemical, physical, and biological agents at work", *Official Journal of the European Communities* (L327), vol. 23, 3 December 1980, pp. 8-13.

Pollution control statutes

Japan

The agencies responsible for environmental protection include the Environmental Protection Agency, the Ministry of International Trade and Industry, the Ministry of Health and Welfare and the Ministry of Agriculture, Forestry and Fisheries. The Environmental Protection Agency has authority over basic policy, general co-ordination of governmental pollution control activities, budgetary policy and research and investigation.

There are no statutes or regulations directed specifically towards the environmental impacts of biotechnology. Thus, the general body of law applies.

The Basic Law for Environmental Pollution Control establishes fundamental national principles and policies and the basic regulatory framework for environmental protection.[85] It empowers the Government to promulgate and enforce environmental quality standards necessary to protect the public health and conserve natural resources.

The basic law is supplemented by laws aimed at specific types of pollution. The Air Pollution Control Law establishes national air quality objectives.[86] The Water Pollution Control Law establishes water quality standards and discharge limits. It also provides for compensation to parties injured by polluted waters or waste products of companies.[87] The Waste Management Law establishes methods for waste disposal.[88] The Chemical Substances Control Law requires manufacturers to test all new chemical substances to be produced in quantities exceeding 100 kilogrammes and to notify the Government of their intent to produce the substance.[89] The Environmental Protection Agency monitors the effects of chemicals in the air and water.

The Laws are supplemented and implemented through Cabinet orders issued by the Prime Minister and through ministerial orders and Environmental Protection Agency notifications. Administrative guidance is used to regulate pollution from specific industrial plants and industries. Local governments also have responsibility, and they may set more stringent standards than those set by the central Government.[90]

United Kingdom

The primary responsibility for the protection of the environment lies with the Department of the Environment, although local governments also have responsibility. In addition, a Royal Commission on Environmental Pollution advises the Government on environmental issues. Although there is no particular legislation or regulations specifically concerned with the environmental impacts of biotechnological products and processes, companies using

---

[85]Basic Law for Environmental Pollution Control, Law No. 132 of 1967, as amended, reprinted in 2 *Int'l. Env't. Rep. Ref. File* (Washington, D.C., Bureau of National Affairs) 91:0501.

[86]Air Pollution Control Law, Law No. 97 of 1968, as amended, reprinted in 2 *Int'l. Env't. Rep. Ref. File* (Washington, D.C., Bureau of National Affairs) 91:0901.

[87]Water Pollution Control Law, Law No. 138 of 1970, as amended, reprinted in 2 *Int'l. Env't. Rep. Ref. File* (Bureau of National Affairs, Washington, D.C.) 91:1401.

[88]Waste Management Law, Law No. 137 of 1970, as amended, reprinted in 2 *Int'l. Env't. Rep. Ref. File* (Bureau of National Affairs, Washington, D.C.) 91:2401.

[89]Chemical Substances Control Law, Law No. 117 of 1973, as amended, reprinted in 2 *Int'l. Env't. Rep. Ref. File* (Bureau of National Affairs, Washington, D.C.) 91:6401.

[90]*Commercial Biotechnology*, p. 558.

biotechnology would be subject to the general environmental laws and regulations. One statute that appears to be particularly relevant to waste products of biotechnological processes is the Controlled Pollution Act of 1974.[91] The act provides for the licensing of sites of disposal of controlled waste, which are defined as household, industrial and commercial waste, both on land and in water. The act was being phased in between July 1983 and July 1986.

United States

There are several statutes dealing with pollution that would apply to biowastes because they generally define pollutants or wastes so as to cover biological materials. They are the Federal Water Pollution Control Act, as amended by the Clean Water Act of 1977;[92] the Marine Protection, Research, and Sanctuaries Act of 1972;[93] the Clean Air Act;[94] and the Solid Waste Disposal Act, as amended by the Resource Conservation and Recovery Act of 1976.[95] These acts prohibit or place restrictions on the discharge of pollutants or waste. Permits are usually required.

Under the Federal Water Pollution Control Act, as amended, the Environmental Protection Agency (EPA) has promulgated regulations governing waste water from the manufacture of pharmaceuticals by fermentation.[96] The regulations place limits on the amount of biological material and solvents that can be discharged.

European Economic Community

EEC has issued no directives or taken any other action specifically to regulate the environmental impact of biotechnology, but several of its general directives concerning waste disposal and water pollution will be applicable to biotechnological products or waste.[97] The regulations are general and flexible, giving maximum discretion and authority to the member States in implementing them.

EEC has a pre-market notification requirement, which is somewhat similar to the Toxic Substances Control Act (TSCA) of the United States, under the

---

[91]Controlled Pollution Act of 1974, reprinted in 3 *Int'l. Env't. Rep. Ref. File* (Bureau of National Affairs, Washington, D.C.) 291:7301.

[92]United States of America 33 U.S.C. 1251-1376 (as amended by Public Law No. 95-217, 91 Stat. 1566 (1977)).

[93]United States of America, 33 U.S.C 1401, 1402, 1411-1421, 1441-1445.

[94]United States of America, 42 U.S.C. 7401-7408, 7521-7574, 7601-7626.

[95]United States of America, 42 U.S.C. 6901-6987 (as amended by Public Law No. 94-580, 90. Stat. 2795 (1976)).

[96]United States of America, 40 C.F.R. Part 439, Subpart A (1985).

[97]Council of the European Communities, "Directive of 4 May 1976, No. 76/464/EEC, on pollution caused by certain dangerous substances discharged in the aquatic environment of the Community", *Official Journal of the European Communities* (L129) vol. 19, 18 May 1976, p. 23, reprinted in 2 *Int.1. Env't. Rep. Ref. File* (Bureau of National Affairs, Washington, D.C.) 151:2101; "Directive of 20 March 1978, No. 78/319/EEC, on toxic and dangerous waste", *Official Journal of the European Communities* (L84) vol. 21, 31 March 1978, p. 43, reprinted in 2 *Int'l. Env't. Rep. Ref. File* (Bureau of National Affairs, Washington, D.C.) 151:1201; "Directive of 17 December 1979, No. 80/684/EEC, on the protection of groundwater against pollution caused by certain dangerous substances", *Official Journal of the European Communities* (L20) vol. 22, 26 January 1980, pp. 43-47, reprinted in 2 *Int'l. Env't. Rep. Ref. File* (Bureau of National Affairs, Washington, D.C.) 151:2101.

sixth amendment to its dangerous substances directive.[98] Under the sixth amendment, a firm must test a new chemical before marketing; it must provide the proper authorities in the member States where the product is to be marketed with the results of the certain minimum testing requirements; and it must conduct any further tests deemed necessary by those authorities before the approval is granted.

Evaluation of current regulation

Several common principles and themes run through the various guidelines and statutes applicable to large-scale fermentation operations. Worker health and safety laws place a general duty on the employer to maintain a safe work-place. The determination of what is a safe work-place often depends on specific standards or requirements, such as those that place limits on exposure to certain substances or agents or that require particular work-place practices. These practices include engineering and quality control, personal protective equipment, warning of hazards, medical surveillance and record keeping. Mechanisms for monitoring and enforcement are provided.

The guidelines for large-scale uses of organisms containing rDNA apply these common principles to a particular situation. They define specific practices and particular containment requirements and procedures for work with rDNA. They also apply the concept of medical surveillance and appropriate oversight.

The many years of experience of pharmaceutical companies with micro-organisms, including some pathogens, and the more recent experience of biotechnology companies with genetically engineered micro-organisms provide a large degree of assurance regarding the safety of large-scale operations. There are some areas of concern, however, especially regarding highly biologically active products. In addition, there are several open issues with respect to the regulation of large-scale processes. One is whether or not guidelines should be developed that encompass other genetic engineering techniques besides rDNA. Another is the type of risk assessment that should be undertaken with respect to worker health and safety. A related issue is what to look for, other than allergic reactions to products, when monitoring workers for potential long-term health impacts of working with genetically engineered organisms.

With respect to biowastes, there appears to be sufficient statutory authority to control such waste in order to prevent adverse impacts on the environment. The application of that authority appears to be fairly straight-forward, since many of the pollution control statutes and certain regulations thereunder specifically mention and deal with biowastes. The controls include government oversight monitoring through requirements for permits and prohibitions or restrictions on discharges. There appears to be a need, however, to consider whether there are any special issues arising from the disposal of genetically engineered organisms and how to deal with those issues within the current regulatory schemes or by way of special guidelines. For example, there

---

[98]Council of the European Communities, "Directive of 18 September 1979, No. 79/831/EEC, amending for the sixth time directive 67/548/EEC on the approximation of the laws, regulations and administrative provisions relating to the classification, packaging and labelling of dangerous substances", *Official Journal of the European Communities*, (L259) vol. 22, 15 October 1979, p. 100.

appears to be a consensus that pathogenic and perhaps all genetically engineered organisms should be killed before being disposed of. Guidelines on how this is done and whether any types of organisms can be exempt probably need to be considered.

In view of the open issues with respect to the safety of large-scale applications, ICGEB can play a major role in risk assessment and the development of international guidelines in this area.

III. Environmental applications[99]

One of the applications of the new biotechnology involves the use of genetically engineered organisms in the environment. Such applications include, for example, plants or animals that have been genetically engineered to enhance one or more desired characteristics, genetically engineered micro-organisms that act as pesticides or deliver agricultural chemicals to plants and genetically engineered micro-organisms that degrade toxic chemicals.

Such applications raise safety issues very different from those raised by the laboratory or factory use of genetically engineered organisms. The critical difference is, of course, the fact that instead of being contained and possibly debilitated, the organisms are intentionally placed in the environment and engineered to be able to survive at least to the extent of doing their intended job, even though they still may be debilitated in comparison to the wild-type organism. Thus, certain safety features employed in the other applications of genetically engineered organisms are not available here. Moreover, there are a number of well-known examples of cases where exotic (non-indigenous) organisms have created adverse or undesirable consequences in new environments. This experience has caused some ecologists to raise concern about the potential risks of genetically engineered organisms that are used in the environment.

On the other hand, other ecologists do not see any special or unique risks raised by environmental uses of genetically engineered organisms. They question the relevance of the experience with exotic organisms and assert that an organism with a few new genes is still essentially the same as the starting organism, which occupies a natural niche in its ecosystem. Moreover, they point to a substantial body of knowledge and experience with a certain type of genetically modified organism, i.e. those created by traditional breeding techniques.

Risk and risk assessment

Three principal and interrelated questions are presented by the use of genetically engineered organisms in the environment. The questions are: what

---

[99]The author has relied mainly on sources from the United States in the present chapter. This is because of the immediacy of this issue in the United States; government policies on this issue appear to be less fully developed in other countries. As with all the sources used in this publication, however, the views of experts are based not only on their own experiences but also on the viewpoints of the international scientific community, many of whose members work permanently or temporarily in the United States or other countries at the forefront of the biotechnological revolution.

risks, if any, are present;[100] whether such risks can be identified and assessed before specific organisms are released; and, if so, how.

Some scientists have raised a number of concerns about the possible adverse consequences of releasing genetically modified organisms into the environment and have suggested that there is insufficient data and experience with such organisms, and with predictive ecology in general, to be able to assess the risk before the organisms are released. In testimony before subcommittees of the United States Congress, ecologist Martin Alexander stated that it was foolhardy to make dogmatic statements about whether or not there will be deleterious effects from genetically engineered organisms in the environment. He also stated that the possibility of harm could not be ruled out at this time and concluded that:

> "It is, thus, my view that alien organisms that are inadvertently or deliberately introduced into natural environments may survive, they may grow, they may find a susceptible host or other environment, and they may do harm. The probability of all these events occurring is small, but the consequences of this low probability event may be enormous.[101]

According to Alexander, because there has not yet been sufficient research or experience with such organisms to assess their risk, "the prudent course of action is to establish the risk factors and simultaneously develop a regulatory procedure to assess survival, growth, and deleterious effects".[101]

Alexander relied on certain arguments to support his conclusion that genetically engineered organisms could have adverse environmental consequences. First, he disputed the notion that all genetic changes in an organism would be disadvantageous to the organism's survival. Secondly, Alexander stated that although most species that are introduced into alien environments do not survive, there are many examples of organisms that do survive and multiply. He cited some well-known cases of disastrous consequences arising from micro-organisms that were introduced into the environment, such as Dutch elm disease in the United States and a fungus that reduced the yield of the corn crop in the United States in 1970 by 10 per cent. Thirdly, Alexander argued that slight changes in the genome of the organisms could alter the harmfulness of those organisms, citing antibiotic resistance in disease-producing micro-organisms, which is often the result of a single gene, and the genetic variant of the influenza virus that appeared in 1918, killing millions of people.[102]

On the basis of testimony presented at this hearing and other information developed by the staff, the Subcommittee on Investigations and Oversight

---

[100]The concern raised about the release of modified micro-organisms relates primarily to potentially adverse environmental consequences, such as harm to desirable plants or animals or overproliferation. Still another area of concern is the remote but possible threat to human health or safety from such organisms.

[101]"Environmental implications of Genetic Engineering: Hearing before the Subcommittee of Investigations and Oversight and the Subcommittee on Science, Research and Technology and the House Committee on Science and Technology", Ninety-eighth Congress, first session 7 (1983) (statement of Martin Alexander, Department of Agronomy, Cornell University, Ithaca, New York).

[102]Ibid. Ecologist Frances E. Sharples, who also testified at the Congressional hearings, agreed with Alexander that only a small fraction of exotic species produced adverse ecological changes but further noted that there was no way to know in advance whether a particular introduced organism would cause disturbances. Sharples observed that ecologists usually did not understand enough about the complex interactions in an ecosystem to be able to predict the outcome of the introduction of a novel organism with any degree of certainty. (Statement of Frances E. Sharples, Oakridge National Laboratory, Oakridge, Tennessee. Ibid., p. 21.

issued a staff report entitled *Environmental Implications of Genetic Engineering*.[103] With respect to the possible risk associated with the release of genetically engineered organisms into the environment, the following conclusion appears in the report:

> "Overall, the risk presented by the deliberate release of a genetically engineered organism is that it may cause environmental changes that perturb the ecosystem it encounters and/or that the organism itself may have negative effects if it establishes itself outside of the specific environment for which it was intended. Although no detrimental effects of any genetically engineered organism on a ecosystem have been documented, severe negative and beneficial *(sic)* impacts from newly introduced naturally occurring organisms are well known. The potential environmental risks associated with the deliberate release of genetically engineered organisms or the translocation of any new organism into an ecosystem are best described as 'low probability, high consequence risk', that is, while there is only a small possibility that damage could occur, the damage that could occur is great."

It should be noted that some scientists have questioned the staff's use of the phrase "high consequence risk" as improperly implying that any adverse consequences would necessarily be severe.

With regard to the current ability to assess the risks presented by such organisms, it was concluded that:

> "The testimony presented to the Subcommittees indicated that predicting the specific type, magnitude, or probability of environmental effects associated with the deliberate release of genetically engineered organisms will be extremely difficult, if not impossible, at the present time. This is principally the case because no historical and scientific data base exists concerning the behavioral characteristics of genetically engineered organisms in the environment and no standard ecological methodology for predicting the outcome of an exotic introduction current *(sic)* exists. In addition, as experiences with naturally occurring organisms have demonstrated, it is possible to make only an imprecise estimate, at best, of the effect that an organism may have on the environment. Nevertheless, the testimony indicated that it would be possible to devise procedures to produce generalized estimates of the probability of environmental damage by, and survival and growth of, a genetically engineered organism, although specific risk assessment may not be achievable."[104]

Other scientists have a quite different perspective on the potential risks. They generally view genetically modified organisms not as "alien" organisms but as mutant organisms similar to those produced by traditional animal and plant breeding techniques. A leading proponent of this view is Winston Brill, a molecular biologist and microbial ecologist. In an article in *Science* on the safety issues raised by agricultural uses of genetically engineered plants and micro-organisms, Brill argued that: *(a)* predictions about the safety of a genetically engineered plant or micro-organisms (one containing rDNA) should be based upon the vast experience with traditional practices, such as plant breeding and the use of microbial soil inoculants; *(b)* an introduced plant or micro-organism containing foreign genes should not be a greater environmental threat than such organisms without recombinant genes; and *(c)* problems caused by the introduction of organisms such as kudzu and the gypsy moth

---

[103] *Staff Report: Environmental Implications of Genetic Engineering*, Subcommittee on Investigations and Oversight, House Committee on Science and Technology, Ninety-eighth Congress, second session 9 (1984).

[104] *Ibid.*, p. 10.

34

into a new environment do not imply problems from an organism, currently considered safe in its habitat, with characterized recombinant genes added to its genome.[105] Brill acknowledged that it may be useful to conduct risk assessment experiments of genetically engineered organisms under appropriate containment.[106]

In support of his arguments, Brill noted that plants have been crossed for centuries to produce new variants, and he asserted that none of the variants had caused serious problems. Such traditional breeding involved the recombination of thousands of genes, and the properties of the progeny were not precisely predictable. Yet breeders had never taken special precautions. Moreover, many genes must interact to cause a problem plant, such as a weed, and Brill suggested that for a major weed such as kudzu,[107] hundreds or possibly thousands of genes must interact. In contrast, recombinant plants would contain a few new genes that would have been well characterized. Brill did not discuss the fact that protoplast fusion of plant cells may mix thousands of uncharacterized genes, however, other than to note that such fusion had produced new variants of plants. Brill made similar arguments for micro-organisms.

That problems encountered in traditional breeding programmes would occur in plant genetic engineering was acknowledged by Brill. For example, plants particularly susceptible to certain pathogens may be inadvertently created, or plants resulting from genetic engineering with uncharacterized genes may contain toxins. He said, however, that there were traditional ways of dealing with those problems.

The United States National Science Foundation (NSF) recently considered the question of how to assess the risks of environmental applications of genetically engineered organisms and issued a report in August 1985.[108] It was concluded in the report that the development of a generic approach to risk assessment was both feasible and desirable. It was further concluded that available risk assessment methods provided a useful foundation for developing a risk assessment approach for those environmental applications. It was noted, however, that only a qualitative approach was currently feasible, given the current state of the art. It was also suggested that several alternative risk assessment approaches were possible. They included deterministic consequence analysis within confidence bounds, qualitative screening and probabilistic risk assessment. The choice of approach would depend on the degree of knowledge about the organism and the corresponding uncertainties about its characteristics under specific environmental conditions. Finally, it was noted that empirical methods, such as microcosm testing, would be indispensable for purposes of risk assessment, but they must be supplemented by predictive modelling methods.

[105]W. J. Brill, "Safety concerns and genetic engineering in agriculture", *Science,* vol. 227, 25 January 1985, p. 381.

[106]Several ecologists have responded to Brill's article in a letter to *Science.* See Colwell and others, *Science* vol. 229, 12 July 1985, pp. 111-115. They argued that Brill's analysis was from the perspective of a geneticist and that as ecologists they would evaluate the potential hazards quite differently on several points. Brill's response to this letter is also published in the same issue, pp. 115-118.

[107]Brill stated that kudzu was a problem because it was introduced into a totally new environment without the usual checks and balances for the plant.

[108]V. T. Covello and J. Fiksel, eds., *The Suitability and Applicability of Risk Assessment Methods for Environmental Applications of Biotechnology* (Washington, D.C., National Science Foundation, 1985).

35

Clearly, there is a scientific controversy over the nature and extent of the risks, if any, presented by the environmental uses of genetically modified organisms. There is also at least some controversy over the extent to which these risks can be assessed, although there appears to be a consensus that the process of hazard identificaton can at least be started and perhaps some qualitative risk assessment can be done. There also seems to be a consensus that any hazard identification or risk assessment will need to be done on a case-by-case basis and that the best way to generate the data will be to proceed with cases.

Regulation of environmental applications

NIH guidelines

Experiments involving the release of organisms containing rDNA into the environment are among those most stringently controlled by the NIH guidelines. Except for certain plants covered by appendix L, these experiments require prior approval by the local IBC, review by RAC and approval by the Director of NIH. Any of these parties can set whatever conditions it considers to be appropriate to protect the environment, including requiring additional data before reviewing the experiment or requiring the organism to be monitored during and after the experiment.

Appendix L applies to experiments involving plants that meet certain conditions. The conditions are: (a) the plant species is a cultivated crop of a genus that has no species known to be a noxious weed; (b) the introduced DNA consists of well-characterized genes containing no sequences harmful to humans, animals or plants; (c) the vector consists of certain specified types of DNA;[109] and (d) the plants are grown in controlled-access fields in conditions appropriate for the plant under study and the geographical location. Experiments meeting these conditions still need prior IBC and NIH approval, but they are reviewed by the RAC Plant Working Group instead of the full RAC and approved by ORDA instead of the Director. RAC has also developed a document covering the desired information to be provided with respect to experiments covered by appendix L (see annex I).

In April 1984, RAC formed a Plant Working Group (PWG) to consider experiments involving the release of organisms containing rDNA into the environment. The Working Group submitted a draft document in which it outlined the information desired in submissions to RAC involving the release of micro-organisms containing rDNA.[110] A copy of the Working Group's guidelines is attached as annex III.

Guidelines of other countries

The Japanese and United Kingdom guidelines require governmental approval before an organism containing rDNA can be released into the

---

[109]The DNA must be from exempt host-vector systems listed in appendix C from plants of the same or closely related species, from nonpathogenic procaryotes or nonpathogenic lower eucaryotic plants or from plant pathogens only if sequences resulting in production of disease symptoms have been deleted, or it must be a chimeric vector constructed from the previously mentioned sequences.

[110]Points to Consider for Submissions Involving Testing in the Environment of Microorganisms Derived by Recombinant DNA Techniques, United States of America, 50 Fed. Reg. 12,456 (1985) (see annex IV).

environment. These countries are in the process of considering the nature and degree of oversight of such activities, but no special guidelines or policy statements for environmental release have been published.[111]

Environmental laws of the United States

The United States has a broad array of environmental laws, which are administered by several agencies. With respect to the environmental use of genetically engineered organisms, the most important agencies are EPA and the United States Department of Agriculture (USDA). All federal agencies, however, must comply with the National Environmental Policy Act (NEPA) before undertaking major actions that significantly affect the environment.[112] The agencies must prepare a written analysis of the adverse environmental effects of such actions and consider alternatives. NEPA has been used by some people and organizations to challenge in court the activities of NIH, EPA and USDA regarding the release of genetically engineered organisms into the environment.[113]

Environmental Protection Agency

Two laws administered by EPA are most relevant to this topic. They are the Federal Insecticide, Fungicide, and Rodenticide Act (FIFRA) and the Toxic Substances Control Act (TSCA).

EPA had detailed its initial policy with regard to applying these two acts to the use of genetically engineered micro-organisms in the environment in the proposal for a co-ordinated framework for regulation of biotechnology, published by the White House's Office of Science and Technology Policy on 31 December 1984.[114] It should be noted that the policy statement is a proposal that represents the initial thinking of EPA. It may be changed in response to comments.

In FIFRA, a pre-marketing clearance procedure is created under which EPA reviews data on a pesticide's safety. It then registers the pesticide for use if it finds that the pesticide will not cause (or significantly increase the risk of) unreasonable adverse effects to humans or the environment when used according to widespread and commonly recognized practice. Labelling requirements and use limitations may be imposed.

In FIFRA, "pesticide" is defined broadly as "any substance or mixture of substances intended for preventing, destroying, repelling, or mitigating any pest" or "any substance or mixture of substances intended for use as a plant regulator, defoilant or dessicant".[115] The term "pest" is also broadly defined to include any insect, rodent, nematode, fungus, weed and virtually any other form of life that the EPA finds to be a pest under certain statutory procedures.[116]

---

[111]An ACGM note on environmental applications was expected by the end of 1986. Telephone interview with M. G. Norton, First Secretary (Science), British Embassy, Washington, D.C., 3 January 1986.

[112]United States of America, 42 U.S.C. 4321-4361.

[113]A discussion of this litigation is beyond the scope of this publication.

[114]"Proposal for a coordinated framework . . .".

[115]United States of America, 7 U.S.C. 136(u).

[116]United States of America, 7 U.S.C. 136(t). Bacteria have been held to be a pest.

Although the definition of pesticide does not specifically mention living organisms, EPA has taken the position for many years that living organisms can be pesticides and, in fact, has registered several micro-organisms for use as pesticides. EPA has recognized, however, that USDA and the Department of the Interior also have regulatory jurisdiction over living organisms. Therefore, it has exempted all organisms from its oversight under FIFRA except for viruses, bacteria, protozoa, fungi and certain unicellular plants.[117] It proposes to continue this exemption for genetically engineered plants and animals.

EPA has proposed a regulation specifying the kinds of data that must be submitted to the Agency to support the registration of a pesticide under FIFRA.[118] Certain sections of the proposed regulation cover biological pest control agents, including genetically engineered ones. These sections set extensive data requirements on product performance, toxicology, residue analysis, hazards to non-target organisms and environmental fate and expression. In addition, EPA has published Pesticide Assessment Guidelines, which specify the standards for conducting acceptable tests, provide guidance on when data are required and on the evaluation and reporting of data and provide examples of the recommended testing protocols.

In a policy statement issued in December 1984, EPA stated that it would require even more information for non-indigenous and genetically engineered microbial pesticides, such as information about the host range and the stability of the organism. EPA considers a non-indigenous organism to be a naturally occurring micro-organism placed in an environment where it is not native. EPA considers the following techniques as coming within the term "genetic engineering": rDNA, rRNA, cell fusion, conjugation, microencapsulation, microinjection, directed or undirected mutagenesis, plasmid transfer and transformation.

Before an unregistered pesticide can be experimentally tested, an experimental use permit from EPA is generally required. The Agency has created an exception from this requirement for tests on less than 10 acres of land or 1 acre of water, if there is no serious environmental or human health concern. Because living organisms multiply and spread beyond the site of application, EPA has reassessed this exception for non-indigenous and genetically engineered microbial pesticides. It has decided to require notification and certain information at least 90 days prior to small-scale field tests so that it may determine if an environmental use permit will be required. This is an interim procedure until a final policy decision is made.[119] On 4 December 1985, EPA announced that it had granted two experimental use permits for the field testing of certain genetically engineered bacteria. The bacteria were designed to control frost damage to strawberries by preventing the colonization of naturally occurring bacteria that produced a protein that caused the formation of ice crystals.[120]

In contrast with FIFRA, TSCA is a notification rather than a pre-market approval act. The burden is on EPA to find a hazard rather than on the manufacturer of the substance to prove safety. TSCA authorizes EPA to acquire information on "chemical substances" in order to identify potential

---

[117]United States of America, 40 C.F.R. 162.5(c)(4).

[118]United States of America, 47 Fed. Reg. 53,192 (1982) (to be codified at 40 C.F.R. Part 158).

[119]"Proposal for a coordinated framework . . .", 50,885.

[120]United States of America, 50 Fed. Reg. 49,761 (1985).

hazards. The act defines a chemical substance as any organic or inorganic substance of a particular molecular identity, including any combination of such substances occurring in whole or in part as a result of a chemical reaction or occurring in nature.[121] EPA can regulate the production, distribution, use and disposal of chemical substances if they present an unreasonable risk of injury to health or the environment. EPA can also require testing of any chemical substance that may present an unreasonable risk of injury to health or the environment or will result in substantial human or environmental exposure. TSCA is intended to fill any gaps in the other environmental statutes.

The heart of TSCA is section 5, in which the goal of the act to identify potentially hazardous new chemical substances before they enter the stream of commerce is implemented. In section 5, manufacturers of any "new chemical substance" are required to notify EPA at least 90 days before beginning manufacture. The notice is called a pre-manufacturing notification (PMN). New chemical substances are chemical substances that are not included on an extensive inventory of existing chemicals compiled by EPA in the late 1970s after TSCA was enacted and updated through the PMN process. Naturally occurring substances are deemed to be on the inventory. The PMNs must disclose known or reasonably ascertainable data about the chemical and its health and environmental effects. EPA has 90 days to prohibit or regulate its distribution; otherwise it is added to the inventory after manufacturing begins.

TSCA also contains reporting and record-keeping requirements useful for information gathering. For example, manufacturers of any chemical substance must maintain records of significant adverse effects and report to EPA information suggesting a substantial risk of injury.

In its policy statement, EPA took the position that TSCA would apply to genetically engineered organisms. Its rationale was that the definition of a chemical substance encompassed nucleic acids and other substances in living organisms and that living organisms were a combination of such substances. Consistent with its position under FIFRA, however, EPA stated it would not apply TSCA to plants and animals.

With respect to micro-organisms, EPA proposed to require PMNs for those produced by rDNA, rRNA and cell fusion. It would not require PMNs for naturally occurring, artifically selected or non-indigenous micro-organisms. It left open the question with respect to those produced by microinjection, microencapsulation, transformation, transduction, transfection, conjugation and plasmid transfer, and undirected mutagenesis. The Agency asked for comments on these.

EPA is considering the applicability of TSCA to the field testing of micro-organisms. TSCA does exempt new chemical substances produced in "small quantities" solely for research and development from the pre-manufacturing notice requirement.[122] EPA is considering the need for prior review of such field tests. It has asserted that it could require such a review by defining "small quantities" so as to exclude such testing from the exemption based on the rationale that living organisms reproduce and spread and therefore, are not tested in small quantities.

---

[121]United States of America, Toxic Substances Control Act, 3(2)(A), 15 U.S.C. 2602(2)(A). The definition excludes chemicals covered by other statutes, such as FIFRA and the Federal Food, Drug, and Cosmetic Act.

[122]United States of America, Toxic Substances Control Act, 5(h)(3), 15 U.S.C. 2604(h)(3).

United States Department of Agriculture

USDA regulates the importation and inter-state shipment of plants, animals and their pathogens under statutes designed to prevent the spread of weeds and plant and animal diseases.[123] This is done by a permit, inspection and quarantine system.

USDA also oversees the introduction of new crop varieties under various mechanisms. These include the National Germ Plasmid Advisory Board and the Plant Variety Protection Office.

In a policy statement issued in December 1984, USDA took the position that genetically modified plants and animals presented no significant new risks when compared to novel, traditionally bred plants and animals and further that its existing regulatory and monitoring authority provided it with sufficient authority to oversee the introduction of genetically engineered plants and animals. USDA apparently would want to oversee the testing of genetically modified organisms and would want to have some information about them, although the policy statement was somewhat vague on this point. It is not yet clear if USDA will require either a pre-testing or pre-marketing review of genetically engineered plants and animals and, if so, what the nature of its review would be.

Environmental statutes of other countries

The environmental laws of Japan and the United Kingdom, which were previously discussed with respect to biowastes, would appear to provide general statutory authority for those governments to control the environmental applications of genetically engineered organisms. There appears to be little in the way of written or published analysis by those countries of how those laws would be applied to this particular situation. This issue appears to be under consideration at this time, and any actions to oversee this application will apparently be in the form of actions under the recombinant DNA guidelines.

Evaluation of current regulation of environmental risks

Countries that exercise broad authority to control risks to the environment would appear to have sufficient general authority to deal with any risks presented by the new biotechnology. That authority includes means for gathering information for risk assessment, various types of notification schemes and prohibitions against certain activities without prior governmental review and approval. Often the authority is implemented by means of a permit.

The critical question is how that authority should be applied to genetic engineering. Several countries are formulating guidelines, policy statements and regulations governing the environmental applications and consequences of genetic engineering. Given the scientific controversy over what risks are presented by the intentional release of genetically engineered organisms and

---

[123]See, in general, United States of America, 21 U.S.C.: 101-135 for animals and their pathogens, 7 U.S.C.; 151-167 for plants; and 7 U.S.C. 150 aa-jj for plant pests. Regulations are found in 9 C.F.R. 71-122.4 (animals and their pathogens) and 7 C.F.R. 300-370.7 (plants and their pathogens).

how to assess those risks, the regulatory process has proceeded at a cautious, deliberate pace. Because there is a fair degree of uncertainty, the authorities are proceeding on a case-by-case basis. It appears, however, that some type of notification and prior approval will be required. Given the current state of affairs in monitoring the environmental aspects of genetic engineering, ICGEB can play a major role in risk assessment and the development of international guidelines.

IV. An international approach to safety issues of genetic engineering

Desirability of an international approach

Individual countries have begun to address the safety issues raised by genetic engineering. It is appropriate for these efforts to continue, since each country will consider the issues from the perspective of its own values and needs. It is also appropriate for international organizations to address these issues from an international perspective, however. Genetic engineering is a technology whose benefits will have major impacts world-wide. Any risks presented by the technology could also have such an impact. International consideration of the issues involving genetic engineering will bring different values and perspectives to bear on them. For example, the concerns of the international community rather than those of particular countries will likely be addressed first. In addition, the needs of the less developed countries would be sure to be considered.

There will be advantages for all countries if an international body or bodies addresses the safety issues of genetic engineering. A major advantage will be a harmonization of regulations on genetic engineering.[124] If there are risks to humans, animals, plants or the environment from genetically engineered organisms that are not properly contained or are not properly evaluated before being deliberately placed into the environment, those risks are not necessarily limited to the country where the organism first entered the environment. Thus, countries can gain a level of security and protection from the careless or reckless conduct of those outside their jurisdiction if there are internationally accepted principles of proper conduct with respect to genetically engineered organisms. Similarly, countries can feel free to develop rational regulation without a fear of driving their biotechnology companies to countries with little or no regulation.

On the other hand, just as the results of international harmonization may place a floor on conduct, they may also place a ceiling on the degree of regulation of genetic engineering. This may be to the benefit of countries whose national regulatory schemes are overly restrictive because of unfounded public concerns rather than careful scientific analysis. It may also prevent the inadvertent creation of trade barriers in such countries to products coming from countries with less restrictive regulation.

Another benefit of harmonization is the fact that it will provide greater certainty and guidance for transnational corporations that are working in many

---

[124]Compare the call of the European Federation of Biotechnology for harmonization in its report, *in* Kuenzi, *op. cit.*

different countries. Otherwise, they might have to face quite different or conflicting regulatory requirements.

Still another advantage of an international approach is the avoidance of a costly duplication of effort in risk assessment and guideline development. This would be especially valuable for countries with limited resources, which could be better directed towards local genetic engineering efforts.

Current international efforts

Efforts have been underway at the international level to identify and assess any risks associated with genetic engineering and to propose guidelines for addressing those risks. WHO is in the process of developing good manufacturing practice guidelines for large-scale processes involving genetically engineered organisms.[125] EEC has also generally considered the risks and regulation of biotechnology but apparently has not begun any efforts to develop guidelines. The European Federation of Biotechnology recently issued a report on the risks of biotechnology.[126] A major effort to consider safety issues was undertaken by the Organization for Economic Co-operation and Development (OECD). In December 1983, OECD created a special *ad hoc* group of government experts on safety and regulation of biotechnology. Under the terms of its mandate, the *ad hoc* group was to review country positions as to the safety in use of genetically engineered organisms at the industrial, agricultural and environmental levels, against the background of existing or planned legislation and regulations for the handling of micro-organisms. In particular, the group was to identify criteria that had been or may be adopted for monitoring the production of genetically engineered organisms and to explore ways of monitoring future production and use of such organisms. The work was seen as a step towards better international harmonization of guidelines and regulations.

The first task of the group was to survey the state of existing regulations in the member countries.[127] The *ad hoc* group produced a summary of the survey in June 1985. Because of difficulties relating to differing definitions of biotechnology in the various countries, the nature of the survey itself and the lack of consistent, detailed responses from all of the countries, however, the summary is only of a very general character.

The *ad hoc* group's second task was to identify scientific criteria or general principles that could serve as the basis of guidelines or regulations.[127] In June 1985, the group produced a draft report entitled "safety and regulations in biotechnology", in which the scientific criteria were considered and standards were set forth that were seen by some as essentially regulatory in nature.[128] The report began with a favourable description of biotechnology and its potential

---

[125]Interview with Vinson R. Oviatt, Director, WHO Special Programme on Safety Measures in Microbiology, Geneva, Switzerland, 9 September 1985.

[126]Kuenzi, *op. cit.*

[127]B. Teso and S. Wald, "Government policy and biotechnology: four key issues", *OECD Observer*, No. 131, November 1984, pp. 16-19.

[128]The basis for the discussion of this draft report is an article that appeared in *Science* in which the draft report and criticisms by some United States experts were discussed. *Science*, vol. 229 (30 August 1985), pp. 842-843. The assertions in the article have generally been confirmed by the author of this publication in discussions with informed and reliable sources in the United States.

applications, followed by a discussion of potential risks, which were generally viewed as minimal. It was also stated, however, that it was impossible to rule out all risk, and, therefore, a general framework to assess the risk and to control genetically engineered organisms used in large-scale processes and for agricultural and environmental purposes was presented.

The document was severely criticized by some experts in the United States on the grounds that it was too pro-regulatory and prescriptive. Instead of being a guideline to conduct, it could become an inflexible standard, according to the critics. Another concern was that the report appeared to be inconsistent with the approach taken by the proposed framework by most of the United States agencies.[129]

The United States delegation to the *ad hoc* working group submitted a proposed revision at a meeting of the group in early December 1985. The group adopted a composite of the June draft and the proposed revision on 5 December 1985. This new document was submitted to the OECD Committee for Scientific and Technological Policy, which approved it on 5 February 1986. The drafting of national guidelines consistent with those of other OECD countries, which had been delayed pending the resolution of the concerns of the United States delegation, is now expected to proceed quickly in Denmark, the Federal Republic of Germany, Japan, the Netherlands and the United Kingdom.[130]

Role of ICGEB and international organizations

Commercial activity in the new biotechnology is increasing significantly, and there is a need to begin comprehensive international efforts in this area quickly. ICGEB and organizations such as UNEP, WHO and UNIDO are playing major but different roles in bringing an international perspective to bear on the safety issues raised by the new biotechnology.

The Centre is the only international scientific institution devoted solely to biotechnology. Member States certainly have an interest in ensuring that activities at the Centre and at affiliated centres are conducted in a safe manner. As a catalyst and role model for many countries interested in developing their biotechnology capabilities, ICGEB will speak and act with institutional authority upon questions relating initially to laboratory safety and later to the safety of large-scale operations and environmental release.

Representatives from international organizations, have already been considering the safety issues of the new biotechnology, and they provided advice with respect to the creation and the activities of ICGEB. They are also likely to be a source of experts for the Centre, and the representatives could act in an informal advisory capacity for the Centre. As United Nations organizations, they have a commonality of interests and perspectives and a commitment to the goals of the Centre. International organizations can continue to assist the Centre in considering and implementing the recommenda-

---

[129]Except for EPA, these agencies take the position that the regulation of genetically engineered products does not present significant and unusual problems and should be addressed by traditional scientific and regulatory principles, whereby the products are evaluated on their own merit and not by the method by which they were produced.

[130]"OECD drafts international guidelines for industry, agriculture, environment", *Biotechnology Newswatch*, 16 December 1985, pp. 2-3.

tions that follow by providing special expertise and acting in an advisory capacity.[131]

UNEP, WHO and UNIDO should continue to be involved in a policy and advisory capacity with these issues because of their special expertise. In addition, the member agencies will continue to be involved individually with biotechnology safety issues because of their organizational mandates. Thus, it would be important to have continuing contacts and co-ordination between agencies and with the Centre.

Recommendations

There are several activities that should be undertaken by the Centre with respect to safety issues raised by genetic engineering. These are the following:

(a) To act as a forum for information exchange and debate;

(b) To study potential risks and publish findings regarding actual hazards and areas where additional research is needed;

(c) To develop risk-assessment methodology;

(d) To conduct risk assessment;

(e) To develop safety guidelines for the various categories of applications of genetically engineered organisms;

(f) To assist countries, especially less developed countries, in adopting the guidelines to their own special needs;

(g) To train scientists, technicians, workers and other support staff to handle genetically engineered organisms and processes involving those organisms safely.

Many of these activities have been initiated by the various international organizations involved in biotechnology. The Centre should take the lead once it is able, and the international organizations should provide expertise, advice and assistance. It should be recognized that the execution of some of these activities would involve the use of consultants or groups of experts.

Forum for information exchange

ICGEB should act as a forum for the exchange of information and debate on biotechnology safety issues. This is a natural consequence of being an international centre and a catalyst for the development of biotechnology.

UNEP, WHO and UNIDO have acted in this capacity and should continue to do so. Even after the Centre initiates its own efforts, these organizations should continue exchanging information about their activities in the area of biotechnology safety and take steps to co-ordinate those activities with each agency and with the Centre. The organizations also could help disseminate information about the work of the Centre beyond the Centre's member States.

---

[131]It may be appropriate and desirable to add representatives from the Food and Agriculture Organization of the United Nations (FAO) and the International Labour Organisation (ILO).

Risk identification and assessment

For ICGEB to be involved with safety issues in a major way, it should identify and study potential risks in order to determine actual hazards. This activity would naturally lead to risk assessment and to further development of risk assessment methodology. It would also lead to the identification of issues in which further research on risk and risk assessment methodology is needed. As experience is gained in assessing the risk, it would be appropriate and desirable for the Centre to conduct risk-assessment experiments, given the scientific facilities and talent of the Centre. The results of such experiments would add to the existing body of knowledge regarding the risks of genetically engineered organisms and would provide additional data on which to base scientifically valid safety guidelines.

Safety guidelines

A major activity for the Informal Working Group already formed between representatives of WHO, UNEP and UNIDO should be to begin work immediately on the safety guidelines that will govern research at the Centre. The group should look to existing statutes and guidelines in the countries that will host ICGEB and affiliated centres for setting initial standards.[132] The Centre, as a result of its deliberations and experience, could modify the initial guidelines. Ultimately, the Centre's guidelines could serve as models for any country where genetic engineering is conducted.

Additional types of guidelines, such as those covering large-scale applications, environmental applications or the handling of biowastes, could be developed in stages by the Centre as it gained experience in those areas and is able to commit the necessary resources. The guidelines should be common, minimum practices that other countries can then implement according to their own special needs. A related role for the Centre would be to assist other countries, especially the less developed countries, in that implementation.

Training

Safety training is an activity most appropriate for the Centre. Training of scientific and technical personnel will be a major function of ICGEB, and the nature of that programme has already been outlined in its basic planning documents.[133] Appropriate safety training should be integrated into this broader programme.

Implementation

ICGEB would be able to implement the programme discussed above in discrete stages with the support and advice of the working group, particularly

---

[132]*Report of the Meeting of the Panel of Scientific Advisors: 11-13 February 1985*, Preparatory Committee on the Establishment of the International Centre for Genetic Engineering and Biotechnology (Sixth session), ICGEB/Prep. Comm./6/9 (New Delhi, India, April 1985).

[133]*Ibid.*, p. 13; "The establishment of an international centre for genetic engineering and biotechnology: report of a group of experts" (UNIDO/IS.254), pp. 15-16.

in the early stages. As the Centre becomes fully operational, the role of the international organizations would probably diminish, although it should not cease because of the valuable expertise and different perspective they can bring to the Centre and its work. One possible way of institutionalizing this role would be to make individual agency representatives *ex officio* advisors to the Centre.

While most and perhaps all of the recommendations could be implemented simultaneously, many would be better implemented in stages after the results of the earlier implemented programmes have been evaluated. For example, it may be appropriate to identify hazards and conduct risk assessment before developing guidelines and to develop guidelines before training personnel in safety practices in other countries. This approach would allow the Centre flexibility and orderly growth. As mentioned previously, however, some guidelines should be in place when ICGEB gets started.

It is suggested that in implementing some of the above recommendations, particularly the development of safety guidelines as models for member and other countries, the Centre should engage in rigorous scientific analysis and procedural safeguards designed to build consensus. Recommendations with regard to risk assessment, guidelines and training should be based on sufficient data, rigorous scientific analysis and evaluation by scientists and policy-makers who represent various interested sectors, such as industry, labour and government. The Centre will be a body with very limited power *vis-à-vis* sovereign state, and its ability to persuade individual countries to follow its examples and recommendations with regard to safety practices will depend ultimately upon the quality of its work and the cogency of its reasoning. Thus, it would be appropriate for the Centre to have a fairly formal mechanism for developing common guidelines for the member countries. By involving scientists and other experts from various countries in the guideline development process, by holding open meetings and by widely circulating drafts to interested parties for comment, the Centre can be assured that an acceptable, consensus document will emerge, even though the process may take longer and involve more effort than other approaches.

For these reasons, it is suggested that the approach of using *ad hoc* groups of experts should be avoided, unless the procedure is consistent with the preceding suggestions. Such groups can lack accountability and can circumvent the consultation process. Efforts in the area of biotechnology safety should be sufficiently formal in structure and sufficiently open to all interested parties to generate confidence in the quality of the finished product.

Annex I

NIH GUIDELINES FOR RESEARCH INVOLVING
RECOMBINANT DNA MOLECULES[a]

DEPARTMENT OF HEALTH AND HUMAN SERVICES

National Institutes of Health

November 1984.

These NIH guidelines supersede earlier versions and will be in effect until further notice.

TABLE OF CONTENTS

---

[a]United States of America, 41 Fed. Reg. 27,902 (1976).

I. Scope of the Guidelines

I-A. *Purpose.* The purpose of these Guidelines is to specify practices for constructing and handling: (i) Recombinant DNA molecules and (ii) organisms and viruses containing recombinant DNA molecules.

I-B. *Definition of Recombinant DNA Molecules.* In the context of these Guidelines, recombinant DNA molecules are defined as either: (i) Molecules which are constructed outside living cells by joining natural or synthetic DNA segments to DNA molecules that can replicate in a living cell, or (ii) DNA molecules that result from the replication of those described in (i) above.

Synthetic DNA segments likely to yield a potentially harmful polynucleotide or polypeptide (e.g., a toxin or a pharmacologically active agent) shall be considered as equivalent to their natural DNA counterpart. If the synthetic DNA segment is not expressed *in vivo* as a biologically active polynucleotide or polypeptide product, it is exempt from the Guidelines.

I-C. *General Applicability.* The Guidelines are applicable to all recombinant DNA research within the United States or its territories which is conducted at or sponsored by an Institution that receives any support for recombinant DNA research from the National Institutes of Health (NIH). This includes research performed by NIH directly.

An individual receiving support for research involving recombinant DNA must be associated with or sponsored by an Institution that can and does assume the responsibilities assigned in these Guidelines.

The Guidelies are also applicable to projects done abroad if they are supported by NIH funds. If the host country, however, has established rules for the conduct of recombinant DNA projects, then a certificate of compliance with those rules may be submitted to NIH in lieu of compliance with the NIH Guidelines. NIH reserves the right to withhold funding if the safety practices to be employed abroad are not reasonably consistent with the NIH Guidelines.

I-D. *General Definitions.* The following terms, which are used throughout the Guidelines, are defined as follows:

I-D-1. *"Institution"* means any public or private entity (including Federal, State, and local government agencies).

I-D-2. "Institutional Biosafety Committee" or "IBC" means a committee that: (i) Meets the requirements for membership specified in Section IV-B-2, and (ii) reviews, approves, and overseas projects in accordance with the responsibilities defined in Sections IV-B-2 and IV-B-3.

I-D-3. "NIH Office of Recombinant DNA Activities" or "ORDA" means the office within NIH with responsibility for: (i) Reviewing and coordinating all activities of NIH related to the Guidelines, and (ii) performing other duties as defined in Section IV-C-3.

I-D-4. "Recombinant DNA Advisory Committee" or "RAC" means the public advisory committee that advises the Secretary, the Assistant Secretary for Health, and the Director of the NIH concerning recombinant DNA research. The RAC shall be constituted as specified in Section IV-C-2.

I-D-5. "Director, NIH" or "Director" means the Director of the NIH or any other officer or employee of NIH to whom authority has been delegated.

II. Containment

Effective biological safety programs have been operative in a variety of laboratories for many years.

Considerable information, therefore, already exists for the design of physical containment facilities and the selection of laboratory procedures applicable to organisms carrying recombinant DNAs (3-16). The existing programs rely upon mechanisms that, for convenience, can be divided into two categories: (i) A set of standard practices that are generally used in microbiological laboratories, and (ii) special procedures, equipment, and laboratory installations that provide physical barriers which are applied in varying degrees according to the estimated biohazard. Four biosafety levels (BL) are described in Appendix G. These biosafety levels consist of combinations of laboratory practices and techniques, safety equipment, and laboratory facilities appropriate for the operations performed and the hazard posed by agents and for the laboratory function and activity. BL4 provides the most stringent containment conditions, BL1 the least stringent.

Experiments on recombinant DNAs by their very nature lend themselves to a third containment mechanism—namely, the application of highly specific biological barriers. In fact, natural barriers do exist which limit either: (i) The effectivity of a *vector* or *vehicle* (plasmid or virus) for specific hosts, or (ii) its dissemination and survival in the environment. The vectors that provide the means for replication of the recombinant DNAs and/or the host cells in which they replicate can be genetically designed to decrease by many orders of magnitude the probability of dissemination of recombinant DNAs outside the laboratory. Further details on biological containment may be found in Appendix I.

As these three means of containment are complementary, different levels of containment appropriate for experiments with different recombinants can be established by applying various combinations of the physical and biological barriers along with a constant use of the standard practices. We consider these categories of containment separately in order that such combinations can be conveniently expressed in the Guidelines.

In constructing these Guidelines, it was necessary to define boundary conditions for the different levels of physical and biological containment and for the classes of experiments to which they apply. We recognize that these definitions do not take into account all existing and anticipated information on special procedures that will allow particular experiments to be carried out under different conditions than indicated here without affecting risk. Indeed, we urge that individual investigators devise simple and more effective containment procedures, and that investigators and IBCs recommend changes in the Guidelines to permit their use.

53

III. Guidelines for Covered Experiments

Part III discusses experiments involving recombinant DNA. These experiments have been divided into four classes:

III-A. Experiments which require specific RAC review and NIH and IBC approval before initiation of the experiment;

III-B. Experiments which require IBC approval before initiation of the experiment;

III-C. Experiments which require IBC notification at the time of initiation of the experiment;

III-D. Experiments which are exempt from the procedures of the Guidelines.

IF AN EXPERIMENT FALLS INTO BOTH CLASS III-A AND ONE OF THE OTHER CLASSES, THE RULES PERTAINING TO CLASS III-A MUST BE FOLLOWED. If an experiment falls into Class III-D and into either Class III-B or III-C as well, it can be considered exempt from the requirements of the Guidelines.

Changes in containment levels from those specified here may not be instituted without the express approval of the Director. NIH (see Sections IV-C-1-b-(1), IV-C-1-b-(2), and subsections).

III-A. *Experiments that Require RAC Review and NIH and IBC Approval Before Initiation.* Experiments in this category cannot be initiated without submission of relevant information on the proposed experiment to NIH, the publication of the proposal in the Federal Register for thirty days of comment, review by the RAC, and specific approval by NIH. The containment conditions for such experiments will be recommended by RAC and set by NIH at the time of approval. Such experiments also require the approval of the IBC before initiation. Specific experiments already approved in this section and the appropriate containment conditions are listed in Appendices D and F. If an experiment is similar to those listed in Appendices D and F, ORDA may determine appropriate containment conditions according to case precedents under Section IV-C-1-b-(3)-(g).

III-A-1. Deliberate formation of recombinant DNAs containing genes for the biosynthesis of toxic molecules lethal for vertebrates at an LD_{50} of less than 100 nanograms per kilogram body weight (e.g., microbial toxins such as the botulinum toxins, tetanus toxin, diptheria toxin, *Shigella dysenteriae* neurotoxin). Specific approval has been given for the cloning in *E. coli* K-12 of DNA containing genes coding for the biosynthesis of toxic molecules which are lethal to vertebrates at 100 nanograms to 100 micrograms per kilogram body weight. Containment levels for these experiments are specified in Appendix F.

III-A-3. Deliberate transfer of a drug resistance trait to microorganisms that are not known to acquire it naturally (2), if such acquisition could compromise the use of the drug to control disease agents in human or veterinary medicine or agriculture.

III-A-4. Deliberate transfer of recombinant DNA or DNA derived from recombinant DNA into human subjects (21). The requirement for RAC review should not be considered to preempt any other required review of experiments with human subjects. Institutional Review Board (IRB) review of the proposal should be completed before submission to NIH.

III-B. *Experiments that Require IBC Approval Before Initiation.* Investigators performing experiments in this category must submit to their IBC, prior to initiation of the experiments, a registration document that contains a description of: (i) The source(s) of DNA; (ii) the nature of the inserted DNA sequences; (iii) the hosts and vectors to be used; (iv) whether a deliberate attempt will be made to obtain expression of a foreign gene, and, if so, what protein will be produced; and (v) the containment conditions specified in these Guidelines. This registration document must be dated and signed by the investigator and filed only with the local IBC. The IBC shall review all such proposals prior to initiation of the experiments. Requests for lowering of containment for experiments in this category will be considered by NIH (see Section IV-C-1-b-(3)).

III-B-1. *Experiments Using Human or Animal Pathogens (Class 2, Class 3, Class 4, or Class 5 Agents (1) as Host-Vector Systems.*

III-B-1-a. Experiments involving the introduction of recombinant DNA into Class 2 agents can be carried out at BL2 containment.

III-B-1-b. Experiments involving the introduction of recombinant DNA into Class 3 agents can be carried out at BL3 containment.

III-B-1-c. Experiments involving the introduction of recombinant DNA into Class 4 agents can be carried out at BL4 containment.

III-B-1-d. Containment conditions for experiments involving the introduction of recombinant DNA into Class 5 agents will be set on a case-by-case basis following ORDA review. A U.S. Department of Agriculture (USDA) permit is required for work with Class 5 agents (18, 20).

III-B-2. *Experiments in which DNA from Human or Animal Pathogens (Class 2, Class 3, Class 4, or Class 5 agents (1) is Cloned in Nonpathogenic Prokaryotic or Lower Eukaryotic Host-Vector Systems.*

III-B-2-a. Recombinant DNA experiments in which DNA from Class 2 or Class 3 agents (1) is transferred into nonpathogenic prokaryotes or lower eukaryotes may be performed under BL2 containment. Recombinant DNA experiments in which DNA from Class 4 agents is transferred into nonpathogenic prokaryotes or lower eukaryotes can be performed at BL2 containment after demonstration that only a totally and irreversibly defective fraction of the agent's genome is present in a given recombinant. In the absence of such a demonstration, BL4 containment should be used. Specific lowering of containment to BL1 for particular experiments can be approved by the IBC. Many experiments in this category will be exempt from the guidelines (see Sections III-D-4 and III-D-5). Experiments involving the formation of recombinant DNAs for certain genes coding for molecules toxic for vertebrates require RAC review and NIH approval (see Section III-A-1) or must be carried out under NIH specified conditions as described in Appendix F.

III-B-2-b. Containment conditions for experiments in which DNA from Class 5 agents is transferred into nonpathogenic prokaryotes or lower eukaryotes will be determined by ORDA following a case-by-case review. A USDA permit is required for work with Class 5 agents (18, 20).

III-B-3. *Experiments Involving the Use of Infectious Animal or Plant Viruses or Defective Animal or Plant Viruses in the Presence of Helper Virus in Tissue Culture Systems.*

Caution: Special care should be used in the evaluation of containment levels for experiments which are likely to either enhance the pathogenicity (e.g., insertion of a host oncogene) or to extend the host range (e.g., introduction of novel control elements) of viral vectors under conditions which permit a productive infection. In such cases, serious consideration should be given to raising the physical containment by at least one level.

Note.—Recombinant DNA molecules which contain less than two-thirds of the genome of any eukaryotic virus (all virus from a single Family (17) being considered identical (19)) may be considered defective and can be used in the absence of helper under the conditions specified in Section III-C.

III-B-3-a. Experiments involving the use of infectious Class 2 animal viruses (1) or defective Class 2 animal viruses in the presence of helper virus can be performed at BL2 containment.

III-B-3-b. Experiments involving the use of infectious Class 3 animal viruses (1) or defective Class 3 animal viruses in the presence of helper virus can be carried out at BL3 containment.

III-B-3-c. Experiments involving the use of infectious Class 4 viruses (1) or defective Class 4 viruses in the presence of helper virus may be carried out under BL4 containment.

III-B-3-d. Experiments involving the use of infectious Class 5 (1) viruses or defective Class 5 viruses in the presence of helper virus will be determined on a case-by-case basis following ORDA review. A USDA permit is required for work with Class 5 pathogens (18, 20).

III-B-3-e. Experiments involving the use of infectious animal or plant viruses or defective animal or plant viruses in the presence of helper virus not covered by Sections III-B-3-a, III-B-3-b, III-B-3-c, or III-B-3-d may be carried out under BL1 containment.

III-B-4. *Recombinant DNA Experiments Involving Whole Animals or Plants.*

III-B-4-a. DNA from any source except for greater than two-thirds of a eukaryotic viral genome may be transferred to any non-human vertebrate organism and propagated under conditions of physical containment comparable to BL1 and appropriate to the organism under study (2). It is important that the investigator demonstrate that the fraction of the viral genome being utilized does not lead to productive infection. A USDA permit is required for work with Class 5 agents (18, 20).

III-B-4-b. For all experiments involving whole animals and plants and not covered by Section III-B-4-a, the appropriate containment will be determined by the IBC (22).

III-B-5. *Experiments Involving More Than 10 Liters of Culture.* The appropriate containment will be decided by the IBC. Where appropriate, Appendix K, *Physical Containment for Large-Scale Uses of Organisms Containing Recombinant DNA Molecules,* should be used.

III-C. *Experiments that Require IBC Notice Simultaneously with Initiation of Experiments.* Experiments not included in Sections III-A, III-B, III-D, and subsections of these sections are to be considered in Section III-C. All such experiments can be carried out at BL1 containment. For experiments in this category, a registration document as described in Section III-B must be dated and signed by the investigator and filed with the local IBC at the time of initiation of the experiment. The IBC shall review all such proposals, but IBC review prior to initiation of the experiment is not required. (The reader should refer to the policy statement in the first two paragraphs of Section IV-A.).

For example, experiments in which all components derive from non-pathogenic prokaryotes and non-pathogenic lower eukaryotes fall under Section III-C and can be carried out at BL1 containment.

Caution: Experiments Involving Formation of Recombinant DNA Molecules Containing no more than Two-Thirds of the Genome of any Eukaryotic Virus. Recombinant DNA molecules containing no more than two-thirds of the genome of any eukaryotic virus (all viruses from a single Family (17) being considered identical (19)) may be propagated and maintained in cells in tissue culture using BL1 containment. For such experiments, it must be shown that the cells lack helper virus for the specific Families of defective viruses being used. If helper virus is presented, procedures specified under Section II-B-3 should be used. The DNA may contain fragments of the genome of viruses from more than one Family but each fragment must be less than two-thirds of a genome.

III-D. *Exempt Experiments.* The following recombinant DNA molecules are exempt from these Guidelines and no registration with the IBC is necessary:

III-D-1. Those that are not in organisms or viruses.

III-D-2. Those that consist entirely of DNA segments from a single nonchromosomal or viral DNA source though one or more of the segments may be a synthetic equivalent.

III-D-3. Those that consist entirely of DNA from a prokaryotic host including its indigenous plasmids or viruses when propagated only in that host (or a closely related strain of the same species) or when transferred to another host by well established physiological means; also, those that consist entirely of DNA from an eukaryotic host including its chloroplasts, mitochondria, or plasmids (but excluding viruses) when propagated only in that host (or a closely related strain of the same species).

III-D-4. Certain specified recombinant DNA molecules that consist entirely of DNA segments from different species that exchange DNA by known physiological processes though one or more of the segments may be a synthetic equivalent. A list of such exchangers will be prepared and periodically revised by the Director, NIH, with advice of the RAC after appropriate notice and opportunity for public comment (see Section IV-C-1-b-(1)-(c)). Certain classes are exempt as of publication of these revised Guidelines. This list is in Appendix A. An updated list may be obtained from the Office of Recombinant DNA Activities, National Institutes of Health, Building 31, Room 3B10, Bethesda, Maryland 20205.

III-D-5. Other classes of recombinant DNA molecules if the Director, NIH, with advice of the RAC, after appropriate notice and opportunity for public comment, finds that they do not present a significant risk to health or the environment (see Section IV-C-1-b-(1)-(c)). Certain classes are exempt as of publication of these revised Guidelines. The list is in Appendix C. An updated list may be obtained from the Office of Recombinant DNA Activities, National Institutes of Health, Building 31, Room 3B10, Bethesda, Maryland 20205.

IV. Roles and Responsibilities

IV-A. *Policy.* Safety in activities involving recombinant DNA depends on the individual conducting them. The Guidelines cannot anticipate every possible situation. Motivation and good judgement are the key essentials to protection of health and the environment.

The Guidelines are intended to help the Institution. Institutional Biosafety Committee (IBC), Biological Safety Officer (BSO), and Principal Investigator (PI) determine the safeguards that should be implemented. These Guidelines will never be complete or final, since all conceivable experiments involving recombinant DNA cannot be foreseen. Therefore, *it is the responsibility of the Institution and those associated with it to adhere to the intent of the Guidelines as well as to their specifics.*

Each Institution (and the IBC acting on its behalf) is responsible for ensuring that recombinant DNA activities comply with the Guidelines. General recognition of institutional authority and responsibility properly establishes accountability for safe conduct of the research at the local level.

The following roles and responsibilities constitute an administrative framework in which safety is an essential and integral part of research involving recombinant DNA molecules. Further clarifications and interpretations of roles and responsibilities will be issued by NIH as necessary.

IV-B. *Responsibilities of the Institution*

IV-B-1. *General Information.* Each Institution conducting or sponsoring recombinant DNA research covered by these Guidelines is responsible for ensuring that the research is carried out in full conformity with the provisions of the Guidelines. In order to fulfil this responsibility, the Institution shall:

IV-B-1-a. Establish and implement policies that provide for the safe conduct of recombinant DNA research and that ensure compliance with the Guidelines. The Institution as part of its general responsibilities for implementing the Guidelines may establish additional procedures as necessary to govern the Institution and its components in the discharge of its responsibilities under the Guidelines. This may include: (i) Statements formulated by the Institution for general implementation of the Guidelines, and (ii) whatever additional precautionary steps the Institution may deem appropriate.

IV-B-1-b. Establish an IBC that meets the requirements set forth in Section IV-B-2 and carries out the functions detailed in Section IV-B-3.

IV-B-1-c. If the Institution is engaged in recombinant DNA research at the BL3 or BL4 containment level, appoint a BSO, who shall be a member of the IBC and carry out the duties specified in Section IV-B-4.

IV-B-1-d. Require that investigators responsible for research covered by these Guidelines comply with the provisions of Section IV-B-5 and assist investigators to do so.

IV-B-1-e. Ensure appropriate training for the IBC chairperson and members, the BSO, PLs, and laboratory staff regarding the Guidelines, their implementation, and laboratory safety. Responsibility for training IBC members may be carried out through the IBC chairperson. Responsibility for training laboratory staff may be carried out through the PI. The Institution is responsible for seeing that the PI has sufficient training but may delegate this responsibility to the IBC.

IV-B-1-f. Determine the necessity in connection with each project for health surveillance of recombinant DNA research personnel, and conduct, if found appropriate, a health surveillance program for the project. (The "Laboratory Safety Monograph" (LSM) discusses various possible components of such a program—for example, records of agents handled, active investigation of relevant illnesses, and the maintenance of serial serum samples for monitoring serologic changes that may result from the employees' work experience. Certain medical conditions may place a laboratory worker at increased risk in any endeavor where infectious agents are handled. Examples given in the LSM include gastrointestinal disorders and treatment with steroids, immunosuppressive drugs, or antibiotics. Workers with such disorders or treatment should be evaluated to determine whether they should be engaged in research with potentially hazardous organisms during their treatment or illness. Copies of the LSM are available from ORDA.)

IV-B-1-g. Report within 30 days to ORDA any significant problems with and violations of the Guidelines and significant research-related accidents and illnesses, unless the institution determines that the PI or IBC has done so.

IV-B-2. *Membership and Procedures of the IBC.* Institution shall establish an IBC whose responsibilities need not be restricted to recombinant DNA. The committee shall meet the following requirements:

IV-B-2-a. The IBC shall comprise no fewer than five members so selected that they collectively have experience and expertise in recombinant DNA technology and the capability to assess the safety of recombinant DNA research experiments and any potential risk to public health or the environment. At least two members shall not be affiliated with the Institution (apart from their membership on the IBC) and shall represent the interest of the surrounding community with respect to health and protection of the environment. Members meet this requirement if, for example, they are officials of State or local public health or environmental protection agencies, members of other local governmental bodies, or persons active in medical, occupational health, or environmental concerns in the community. The BSO, mandatory when research is being conducted at the BL3 and BL4 levels, shall be a member (see Section IV-B-4).

IV-B-2-b. In order to ensure the competence necessary to review recombinant DNA activities, it is recommended that: (i) The IBC include persons with expertise in recombinant DNA technology, biological safety, and physical containment; (ii) the IBC include, or have available as consultants, persons knowledgeable in institutional commitments and policies, applicable law, standards of professional conduct and practice, community attitudes, and the environment; and (iii) at least one member be from the laboratory technical staff.

IV-B-2-c. The Institution shall identify the committee members by name in a report to ORDA and shall include relevant background information on each member in such form and at such times as ORDA may require.

IV-B-2-d. No member of an IBC may be involved (except to provide information requested by the IBC) in the review or approval of a project in which he or she has been or expects to be engaged or has a direct financial interest.

IV-B-2-e. The Institution, who is ultimately responsible for the effectiveness of the IBC, may establish procedures that the IBC will follow in its initial and continuing review of applications, proposals, and activities. (IBC review procedures are specified in Section IV-B-3-a.)

IV-B-2-f. Institutions are encouraged to open IBC meetings to the public whenever possible, consistent with protection of privacy and proprietary interests.

IV-B-2-g. Upon request, the Institution shall make available to the public all minutes of IBC meetings and any documents submitted to or received from funding agencies which the latter are required to make available to the public. If comments are made by members of the public on IBC actions, the Institution shall forward to NIH both the comments and the IBC's response.

IV-B-3. *Functions of the IBC.* On behalf of the Institution, the IBC is responsible for:

IV-B-3-a. Reviewing for compliance with the NIH Guidelines recombinant DNA research as specified in Part III conducted at or sponsored by the Institution, and approving those research projects that it finds are in conformity with the Guidelines. This review shall include:

IV-B-3-a-(1). An independent assessment of the containment levels required by these Guidelines for the proposed research, and

IV-B-3-a-(2). An assessment of the facilities, procedures, and practices, and of the training and expertise of recombinant DNA personnel.

IV-B-3-b. Notifying the PI of the results of their review.

IV-B-3-c. Lowering containment levels for certain experiments as specified in Section III-B-2.

IV-B-3-d. Setting containment levels as specified in Section III-B-4-d and III-B-5.

IV-B-3-e. Reviewing periodically recombinant DNA research being conducted at the Institution to ensure that the requirements of the Guidelines are being fulfilled.

IV-B-2-f. Adopting emergency plans covering accidental spills and personnel contamination resulting from such research.

Note.—Basic elements in developing specific procedures for dealing with major spills of potentially hazardous materials in the laboratory are detailed in the LSM. Included are information and references on decontamination and emergency plans. NIH and the Centers for Disease Control are available to provide consultation and direct assistance, if necessary, as posted in the LSM. The Institution shall co-operate with the State and local public health departments reporting any significant research-related illness or accident that appears to be a hazard to the public health.

IV-B-3-g. Reporting within 30 days to the appropriate institutional official and to ORDA any significant problems with or violations of the Guidelines and any significant research-related accidents or illnesses unless the IBC determines that the PI has done so.

IV-B-3-h. The IBC may not authorize initiation of experiments not explicitly covered by the Guidelines until NIH (with the advice of the RAC when required) established the containment requirement.

IV-B-3-i. Performing such other functions as may be delegated to the IBC under Section IV-B-1.

IV-B-4. *Biological Safety Officer.* The Institution shall appoint a BSO if it engages in recombinant DNA research at the BL3 or BL4 containment level. The officer shall be a member of the IBC, and his or her duties shall include (but need not be limited to):

IV-B-4-a. Ensuring through periodic inspections that laboratory standards are rigorously followed;

IV-B-4-b. Reporting to the IBC and the Institution all significant problems with and violations of the Guidelines and all significant research-related accidents and illnesses of which the BSO becomes aware unless the BSO determines that the PI has done so;

IV-B-4-c. Developing emergency plans for dealing with accidental spills and personnel contamination and investigating recombinant DNA research laboratory accidents;

IV-B-4-d. Providing advice on laboratory security;

IV-B-4-e. Providing technical advice to the PI and the IBC on research safety procedures.

Note.—See LSM for additional information on the duties of the BSO.

IV-B-5. *Principal Investigator (PI).* On behalf of the Institution, the PI is responsible for complying fully with the Guidelines in conducting any recombinant DNA research.

IV-B-5-a. PI—*General.* As part of this general responsibility, the PI shall:

IV-B-5-a-(1). Initiate or modify no recombinant DNA research requiring approval by the IBC prior to initiation (see Sections III-A and III-B) until that research or the proposed modification thereof has been approved by the IBC and has met all other requirements of the Guidelines;

IV-B-5-a-(2). Determine whether experiments are covered by Section III-C and follow the appropriate procedures;

IV-B-5-a-(3). Report within 30 days to the IBC and NIH (ORDA) all significant problems with and violations of the Guidelines and all significant research-related accidents and illnesses;

IV-B-5-a-(4).   Report to the IBC and to NIH (ORDA) new information bearing on the Guidelines;

IV-B-5-a-(5).   Be adequately trained in good microbiological techniques;

IV-B-5-a-(6).   Adhere to IBC-approved emergency plans for dealing with accidential spills and personal contamination; and

IV-B-5-a-(7).   Comply with shipping requirements for recombinant DNA molecules. (See Appendix H for shipping requirements and the LSM for technical recommendations.)

IV-B-5-b.   *Submissions by the PI to NIH.*   The PI shall:

IV-B-5-b-(1).   Submit information to NIH (ORDA) in order to have new host-vector systems certified;

IV-B-5-b-(2).   Petition NIH with notice to the IBC for exemptions to these Guidelines;

IV-B-5-b-(3).   Petition NIH with concurrence of the IBC for approval to conduct experiments specified in Section III-A of the Guidelines;

IV-B-5-b-(4).   Petition NIH for determination of containment for experiments requiring case-by-case review;

IV-B-5-b-(5).   Petition NIH for determination of containment for experiments not covered by the Guidelines.

IV-B-5-c.   *Submissions by the PI to the IBC.*   The PI shall:

IV-B-5-c-(1).   Make the initial determination of the required levels of physical and biological containment in accordance with the Guidelines;

IV-B-5-c-(2).   Select appropriate microbiological practices and laboratory techniques to be used in the research;

IV-B-5-c-(3).   Submit the initial research protocol if covered under Guidelines Sections III-A, III-B, or III-C (and also subsequent changes—e.g., changes in the source of DNA or host-vector system) to the IBC for review and approval or disapproval; and

IV-B-5-c-(4).   Remain in communication with the IBC throughout the conduct of the project.

IV-B-5-d.   *PI Responsibilities Prior to Initiating Research.*   The PI is responsible for:

IV-B-5-d-(1).   Making available to the laboratory staff copies of the protocols that describe the potential biohazards and the precautions to be taken;

IV-B-5-d-(2).   Instructing and training staff in the practices and techniques required to ensure safety and in the procedures for dealing with accidents; and

IV-B-5-d-(3).   Informing the staff of the reasons and provisions for any precautionary medical practices advised or requested, such as vaccinations or serum collection.

IV-B-5-e.   *PI Responsibilities During the Conduct of the Research.*   The PI is responsible for:

IV-B-5-e-(1).   Supervising the safety performance of the staff to ensure that the required safety practices and techniques are employed;

IV-B-5-e-(2). Investigating and reporting in writing to ORDA, the BSO (where applicable), and the IBC any significant problems pertaining to the operation and implementation of containment practices and procedures;

IV-B-5-e-(3). Correcting work errors and conditions that may result in the release of recombinant DNA materials;

IV-B-5-e-(4). Ensuring the integrity of the physical containment (e.g., biological safety cabinets) and the biological containment (e.g., purity and genotypic and phenotypic characteristcs).

IV-C. *Responsibilities of NIH.*

IV-C-1. *Director.* The Director, NIH, is responsible for: (i) Establishment of the NIH Guidelines for Research Involving Recombinant DNA Molecules, (ii) overseeing their implementation, and (iii) their final interpretation.

The Director has responsibilities under the Guidelines that involve ORDA and RAC. ORDA's responsibilities under the Guidelines are administrative. Advice from the RAC is primarily scientific and technical. In certain circumstances, there is specific opportunity for public comment with published response before final action.

IV-C-1-a. *General Responsibilities of the Director, NIH.* The responsibilities of the Director shall include the following:

IV-C-1-a-(1). Promulgating requirements as necessary to implement the Guidelines;

IV-C-1-a-(2). Establishing and maintaining the RAC to carry out the responsibilities set forth in Section IV-C-2. The RAC's membership is specified in its charter and in Section IV-C-2;

IV-C-1-a-(3). Establishing and maintaining ORDA to carry out the responsibilities defined in Section IV-C-3; and

IV-C-1-a-(4). Maintaining the Federal Interagency Advisory Committee on Recombinant DNA Research established by the Secretary, HEW (now HHS), for advice on the coordination of all Federal programs and activities relating to recombinant DNA including activities of the RAC (see Appendix J).

IV-C-1-b. *Specific Responsibilities of the Director, NIH.* In carrying out the responsibilities set forth in this section, the Director or a designee shall weigh each proposed action through appropriate analysis and consultation to determine that it complies with the Guidelines and presents no significant risk to health or the environment.

IV-C-1-b-(1). *Major Actions.* To execute major actions the Director must seek the advice of the RAC and provide an opportunity for public and Federal agency comment. Specifically, the agenda of the RAC meeting citing the major actions will be published in the Federal Register at least 30 days before the meeting, and the Director will also publish the proposed actions in the Federal Register for comment at least 30 days before the meeting. In addition, the Director's proposed decision, at his discretion, may be published in the Federal Register for 30 days of comment before final action is taken. The Director's final decision, along with response to the comments, will be published in the Federal Register and the *Recombinant DNA Technical Bulletin.* The RAC and IBC chairpersons will be notified of this decision:

IV-C-1-b-(1)-(a). Changing containment levels for types of experiments that are specified in the Guidelines when a major action is involved;

IV-C-1-b-(1)-(b). Assigning containment levels for types of experiments that are not explicity considered in the Guidelines when a major action is involved;

IV-C-1-b-(1)-(c). Promulgating and amending a list of classes of recombinant DNA molecules to be exempt from these Guidelines because they consist entirely of DNA segments from species that exchange DNA by known physiological processes or otherwise do not present a significant risk to health or the environment;

IV-C-1-b-(1)-(d). Permitting experiments specified by Section III-A of the Guidelines;

IV-C-1-b-(1)-(e). Certifying new host-vector systems with the exception of minor modifications of already certified systems (the standards and procedures for certification are described in Appendix I-II-A. Minor modifications constitute, for example, those of minimal or no consequence to the properties relevant to containment); and

IV-C-1-b-(1)-(f). Adopting other changes in the Guidelines.

IV-C-1-b-(2). *Lesser Actions.* To execute lesser actions, the Director must seek the advice of the RAC. The Director's decision will be transmitted to the RAC and IBC chairpersons and published in the *Recombinant DNA Technical Bulletin:*

IV-C-1-b-(2)-(a). Interpreting and determining containment levels upon request by ORDA;

IV-C-1-b-(2)-(b). Changing containment levels for experiments that are specified in the Guidelines (see Section III);

IV-C-1-b-(2)-(c). Assigning containment levels for experiments not explicitly considered in the Guidelines;

IV-C-1-b-(2)-(d). Revising the "Classification of Etiologic Agents" for the purpose of these Guidelines (1).

IV-C-1-b-(3). *Other Actions.* The Director's decision will be transmitted to the RAC and IBC chairpersons and published in the *Recombinant DNA Technical Bulletin:*

IV-C-1-b-(3)-(a). Interpreting the Guidelines for experiments to which the Guidelines specifically assign containment levels;

IV-C-1-b-(3)-(b). Setting containment under Section III-B-1-d and Section III-B-3-d;

IV-C-1-b-(3)-(c). Approving minor modifications of already certified host-vector systems (the standards and procedures for such modifications are described in Appendix I-II);

IV-C-1-b-(3)-(d). Decertifying already certified host-vector systems;

IV-C-1-b-(3)-(e). Adding new entries to the list of molecules toxic for vertebrates (see Appendix F);

IV-C-1-b-(3)-(f). Approving the cloning of toxin genes in host-vector systems other than *E. coli* K-12 (see Appendix F); and

IV-C-1-b-(3)-(g). Determining appropriate containment conditions for experiments according to case precedents developed under Section IV-C-1-b-(2)-(c).

IV-C-1-b-(4). The Director shall conduct, support, and assist training programs in laboratory safety for IBC members, BSOs, PIs, and laboratory staff.

IV-C-2. *Recombinant DNA Advisory Committee.* The Recombinant DNA Advisory Committee (RAC) is responsible for carrying out specified functions cited below as well as others assigned under its charter or by the Secretary, HHS, the Assistant Secretary for Health, and the Director, NIH.

The committee shall consist of 25 members, including the chair, appointed by the Secretary or designee, at least fourteen of whom shall be selected from authorities knowledgeable in the fields of molecular biology or combinant DNA research or in scientific fields other than molecular biology or recombinant DNA research, and at least six of whom shall be persons knowledgeable in applicable law, standards of professional conduct and practice, public attitudes, the environment, public health, occupational health, or related fields. Representatives from Federal agencies shall serve as non-voting members Nominations for the RAC may be submitted to the Office of Recombinant DNA Activities, National Institutes of Health, Building 31, Room 3B10, Bethesda, MD 20205.

All meetings of the RAC will be announced in the Federal Register, including tentative agenda items, 30 days in advance of the meeting with final agendas (if modified) available at least 72 hours before the meeting. No items defined as a major action under Section IV-C-1-b-(1) may be added to an agenda after it appears in the Federal Register.

The RAC shall be responsible for advising the Director, NIH, on the actions listed in Section IV-C-1-b-(1) and IV-C-1-b-(2).

IV-C-3. *The Office of Recombinant DNA Activities.* ORDA shall serve as a focal point for information on recombinant DNA activities and provide advice to all within and outside NIH including Institutions, BSOs, PIs, Federal agencies, State and local governments and institutions in the private sector. ORDA shall carry out such other functions as may be delegated to it by the Director, NIH, including those authorities described in Section IV-C-1-b-(3). In addition, ORDA shall be responsible for the following:

IV-C-3-a. Reviewing and approving IBC membership;

IV-C-3-b. Publishing in the Federal Register:

IV-C-3-b-(1). Announcements of RAC meetings and agendas at least 30 days in advance;

Note.—If the agenda for an RAC meeting is modified, ORDA shall make the revised agenda available to anyone upon request at least 72 hours in advance of the meeting.

IV-C-3-b-(2). Proposed major actions of the type falling under Section IV-C-1-b-(1) at least 30 days prior to the RAC meeting at which they will be considered; and

IV-C-3-b-(3). The NIH Director's final decision on recommendations made by the RAC.

IV-C-3-c. Publishing the *Recombinant DNA Technical Bulletin;* and

IV-C-3-d. Serving as executive secretary of the RAC.

IV-C-4. *Other NIH Components.* Other NIH components shall be responsible for certifying maximum containment (BL4) facilities, inspecting them periodically, and inspecting other recombinant DNA facilities as deemed necessary.

IV-D. *Compliance.* As a condition for NIH funding of recombinant DNA research, Institutions must ensure that such research conducted at or sponsored by the Institution, irrespective of the source of funding, shall comply with these Guidelines. The policies on noncompliance are as follows:

IV-D-1. All NIH-funded projects involving recombinant DNA techniques must comply with the NIH Guidelines. Noncompliance may result in: (i) Suspension, limitation, or termination of financial assistance for such projects and of NIH funds for other recombinant DNA research at the Institution, or (ii) a requirement for prior NIH approval of any or all recombinant DNA projects at the Institution.

IV-D-2. All non-NIH funded projects involving recombinant DNA techniques conducted at or sponsored by an Institution that receives NIH funds for projects involving such techniques must comply with the NIH Guidelines. Noncompliance may result in: (i) Suspension, limitation, or termination of NIH funds for recombinant DNA research at the Institution, or (ii) a requirement for prior NIH approval of any or all recombinant DNA projects at the Institution.

IV-D-3. Information concerning noncompliance with the Guidelines may be brought forward by any person. It should be delivered to both NIH (ORDA) and the relevant Institution. The Institution, generally through the IBC, shall take appropriate action. The Institution shall forward a complete report of the incident to ORDA, recommending any further action.

IV-D-4. In cases where NIH proposes to suspend, limit, or terminate financial assistance because of noncompliance with the Guidelines, applicable DHHS and Public Health Service procedures shall govern.

See Appendix J for information on the Federal Interagency Advisory Committee on Recombinant DNA Research.

IV-D-5. *Voluntary Compliance.* Any individual, corporation, or institution that is not otherwise covered by the Guidelines is encouraged to conduct recombinant DNA research activities in accordance with the Guidelines through the procedures set forth in Part VI.

V. Footnotes and References of Sections I-IV

(1) The original reference to organisms as Class 1, 2, 3, 4, or 5 refers to the classification in the publication *Classification of Etiologic Agents on the Basis of Hazard,* 4th Edition, July 1974: U.S. Department of Health, Education, and Welfare, Public Health Service, Centers for Disease Control, Office of Biosafety, Atlanta, Georgia 30333.

The Director, NIH, with advice of the Recombinant DNA Advisory Committee, may revise the classification for the purposes of these Guidelines (see Section IV-C-1-b-(2)-(d)). The revised list of organisms in each class is reprinted in Appendix B to these Guidelines.

(2) In Part III of the Guidelines, there are a number of places where judgments are to be made. In all these cases the principal investigator is to make the judgment on these matters as part of his responsibility to "make the initial determination of the required levels of physical and biological containment in accordance with the Guidelines" (Section IV-B-5-c-(1)). In the cases falling under Sections III-A, -B or -C, this judgment is to be reviewed and approved by the IBC as part of its responsibility to make "an independent assessment of the containment levels required by these Guidelines for the proposed research" (Section IV-B-3-a-(1)). If the IBC wishes, any specific cases may be referred to ORDA as part of ORDA's functions to "provide advice to all within and outside NIH" (Section IV-C-3), and ORDA may request advice from the RAC as part of the RAC's responsibility for "interpreting and determining containment levels upon request by ORDA" (Section IV-C-1-b-(2)-(a)).

(3) *Laboratory Safety at the Center for Disease Control* (Sept. 1974). U.S. Department of Health, Education and Welfare Publication No. CDC 75-8118.

(4) *Classification of Etiologic Agents on the Basis of Hazard* (4th Edition, July 1974). U.S. Department of Health, Education and Welfare. Public Health Service, Centers for Disease Control, Office of Biosafety, Atlanta, Georgia 30333.

(5) *National Cancer Institute Safety Standards for Research Involving Oncogenic Viruses* (Oct. 1974), U.S. Department of Health, Education and Welfare Publication No. (NIH) 75-790.

(6) *National Institutes of Health Biohazards Safety Guide* (1974). U.S. Department of Health, Education, and Welfare, Public Health Service, National Institutes of Health, U.S. Government Printing Office, Stock No. 1740-00383.

(7) *Biohazards in Biological Research (1973).* A. Hellman, M. N. Oxman, and R. Pollack (ed.) Cold Spring Harbor Laboratory.

(8) *Handbook of Laboratory Safety* (1971). 2nd Edition, N. V. Steere (ed.). The Chemical Rubber Co., Cleveland.

(9) Bodily, J. L. (1970). *General Administration of the Laboratory,* H. L. Bodily, E. L. Updyke, and J. O. Mason (eds.). Diagnostic Procedures for Bacterial, Mycotic and Parasitic Infections. American Public Health Association, New York, pp. 11-28.

(10) Darlow, H. M. (1969). *Safety in the Microbiological Laboratory.* In J. R. Norris and D. W. Robbins (ed.), Methods in Microbiology. Academic Press, Inc., New York, pp. 169-204.

(11) *The Prevention of Laboratory Acquired Infection* (1974). C. H. Collins, E. G. Hartley, and R. Pilsworth. Public Health Laboratory Service, Monograph Series No. 6.

(12) Chatigny, M. A. (1961). *Protection Against Infection in the Microbiological Laboratory: Devices and Procedures.* In W. W. Umbreit (ed.). Advances in Applied Microbiology. Academic Press, New York, N.Y. 3:131-192.

(13) *Design Criteria for Viral Oncology Research Facilities* (1975). U.S. Department of Health, Education and Welfare. Public Health Service, National Institutes of Health, DHEW Publication No. (NIH) 75-891.

(14) Kuehne, R. W. (1973). *Biological Containment Facility for Studying Infectious Disease.* Appl. Microbiol. 28-239-243.

(15) Runkle, R. S., and G. B. Phillips (1969). *Microbial Containment Control Facilities.* Van Nostrand Reinhold, New York.

(16) Catigny, M. A., and D. I. Clinger (1969). *Contamination Control in Aerobiology.* In R. L. Dimmick, and A. B. Akers (eds.). An Introduction to Experimental Aerobiology. John Wiley & Sons, New York, pp. 194-263.

(17) As classified in the Third Report of the International Committee on Taxonomy of Viruses: Classification and Nomenclature of Viruses, R. E. F. Matthews, Ed. Intervirology 12 (129-296) 1979.

(18) A USDA permit, required for import and interstate transport of pathogens, may be obtained from the Animal and Plant Health Inspection Service, USDA, Federal Building, Hyattsville, MD 20782.

(19) i.e., the total of all genomes within a Family shall not exceed two-thirds of the genome.

(20) All activities, including storage of variola and whitepax, are restricted to the single national facility (World Health Organization (WHO) Collaborating Center for Smallpox Research, Centers for Disease Control, in Atlanta).

(21) Section III-A-4 covers only those experiments in which the intent is to modify stably the genome of cells of a human subject. Other experiments involving recombinant DNA in human subjects such as feeding of bacteria containing recombinant DNA or the administration of vaccines containing recombinant DNA are not covered in Section III-A-4 of the Guidelines.

(22) For recombinant DNA experiments in which the intent is to modify stably the genome of cells of a human subject, see Section III-A-4.

VI. Voluntary Compliance

VI-A. *Basic Policy.* Individuals, corporations, and institutions not otherwise covered by the Guidelines are encouraged to do so by following the standards and procedures set forth in Parts I-IV of the Guidelines. In order to simplify discussion, references hereafter to "institutions" are intended to encompass corporations, and individuals who have no organizational affiliation. For purposes of complying with the Guidelines, an individual intending to carry out research involving recombinant DNA is encouraged to affiliate with an institution that has an IBC approved under the Guidelines.

Since commercial organizations have special concerns, such as protection of proprietary data, some modifications and explanations of the procedures in Parts I-IV are provided below, in order to address these concerns.

66

VI-B. *IBC Approval.* ORDA will review the membership of an institution's IBC and where it finds the IBC meets the requirements set forth in Section IV-B-2 will give its approval to the IBC membership.

It should be emphasized that employment of an IBC member solely for purposes of membership on the IBC does not itself make the member an institutionally affiliated member for purposes of Section IV-B-2-a.

Except for the unaffiliated members, a member of an IBC for an institution not otherwise covered by the Guidelines may participate in the review and approval of a project in which the member has a direct financial interest so long, as the member has not been and does not expect to be engaged in the project. Section IV-B-2-d is modified to that extent for purposes of these institutions.

VI-C. *Certification of Host-Vector Systems.* A host-vector system may be proposed for certification by the Director, NIH, in accordance with the procedures set forth in Appendix I-II-A.

In order to ensure protection for proprietary data, any public notice regarding a host-vector system which is designated by the institution as proprietary under Section VI-E-1 will be issued only after consultation with the institution as to the content of the notice.

VI-D. *Requests for Exemptions and Approvals.* Requests for exemptions or other approvals required by the Guidelines should be requested by following the procedures set forth in the appropriate sections in Parts I-IV of the Guidelines.

In order to ensure protection for proprietary data, any public notice regarding a request for an exemption or other approval which is designated by the institution as proprietary under Section VI-E-1 will be issued only after consultation with the institution as to the content of the notice.

VI-E. *Protection of Proprietary Data.* In general, the Freedom of Information Act requires Federal agencies to make their records available to the public upon request. However, this requirement does not apply to, among other things, "trade secrets and commercial and financial information obtained from a person and privileged or confidential." 18 U.S.C. 1905, in turn makes it a crime for an officer or employee of the United States or any Federal department or agency to publish, divulge, disclose, or make known "in any manner or to any extent not authorized by law any information coming to him in the course of his employment or official duties or by reason of any examination or investigation made by, or return, report or record made to or filed with, such department or agency or officer or employee thereof, which information concerns or relates to the trade secrets, (or) processes . . . of any person, firm, partnership, corporation, or association." This provision applies to all employees of the Federal Government, including special Government employees. Members of the Recombinant DNA Advisory Committee are "special Government employees."

VI-E-1. In submitting information to NIH for purposes of complying voluntarily with the Guidelines, an institution may designate those items of information which the institution believes constitute trade secrets, privileged, confidential commercial, or financial information.

VI-E-2. If NIH receives a request under the Freedom of Information Act for information so designated, NIH will promptly contact the institution to secure its views as to whether the information (or some portion) should be released.

VI-E-3. If the NIH decides to release this information (or some portion) in response to a Freedom of Information request or otherwise, the institution will be advised; and the actual release will not be made until the expiration of 15 days after the institution is so advised except to the extent that earlier release in the judgment of the Director, NIH, is necessary to protect against an imminent hazard to the public or the environment.

VI-E-4. *Presubmission Review.*

VI-E-4-a. Any institution not otherwise covered by the Guidelines, which is considering submission of data or information voluntarily to NIH, may request presubmission review of the records involved to determine whether the records are submitted NIH will or will not make part or all of the records available upon request under the Freedom of Information Act.

VI-E-4-b. A request for presubmission review should be submitted to ORDA along with the records involved. These records must be clearly marked as being the property of the institution on loan to NIH solely for the purpose of making a determination under the Freedom of Information Act. ORDA will then seek a determination from the HHS Freedom of Information Officer, the responsible official under HHS regulations (45 C.F.R. Part 5) as to whether the records involved (or some portion) are or are not available to members of the public under the Freedom of Information Act. Pending such a determination the records will be kept separate from ORDA files, will be considered records of the institution and not ORDA, and will not be received as part of ORDA files. No copies will be made of the records.

VI-E-4-c. ORDA will inform the institution of the HHS Freedom of Information Officer's determination and follow the institution's instructions as to whether some or all of the records involved are to be returned to the institution or to become a part of ORDA files. If the institution instructs ORDA to return the records, no copies or summaries of the records will be made or retained by HHS, NIH, or ORDA.

VI-E-4-d. The HHS Freedom of Information Officer's determination will represent that official's judgement at the time of the determination as to whether the records involved (or some portion) would be exempt from disclosure under the Freedom of Information Act if at the time of the determination the records were in ORDA files and a request were received for them under the Act.

Appendix A.   Exemptions Under Section III-D-4

Section III-D-4 states that exempt from these Guidelines are "certain specified recombinant DNA molecules that consist entirely of DNA segments from different species that exchange DNA by known physiological processes though one or more of the segments may be a synthetic equivalent. A list of such exchangers will be prepared and periodically revised by the Director, NIH, with advice of the RAC after appropriate notice and opportunity for public comment (see Section IV-C-1-b-(1)-(c)). Certain classes are exempt as of publication of these revised Guidelines. The list is in Appendix A."

Under Section III-D-4 of these Guidelines are recombinant DNA molecules that are: (1) composed entirely of DNA segments from one or more of the organisms within a sublist and (2) to be propagated in any of the organisms within a sublist. (Classification of *Bergey's Manual of Determinative Bacteriology,* 8th edition. R. E. Buchanan and N. E. Gibbons, editors. Williams and Wilkins Company: Baltimore, 1974.)

Sublist A

1. Genus *Escherichia*
2. Genus *Shigella*
3. Genus *Salmonella* (including *Arizona*)
4. Genus *Enterobacter*
5. Genus *Citrobacter* (including *Levinea*)
6. Genus *Klebsiella*

7. Genus *Erwinia*
8. *Pseudomonas aeruginosa, Pseudomonas putida* and *Pseudomonas fluorescens*
9. *Serratia marcescens*
10. *Yersinia enterocolitica*

Sublist B

1. *Bacillus subtilis*
2. *Bacillus licheniformis*
3. *Bacillus pumilus*
4. *Bacillus globigii*
5. *Bacillus niger*
6. *Bacillus nato*
7. *Bacillus amyloliguefaciens*
8. *Bacillus aterrimus*

Sublist C

1. *Streptomyces aureofaciens*
2. *Streptomyces rimosus*
3. *Streptomyces coelicolor*

Sublist D

1. *Streptomyces griseus*
2. *Streptomyces cyaneus*
3. *Streptomyces venezuelae*

Sublist E

1. One way transfer of *Streptococcus mutans* or *Streptococcus lactis* DNA into *Streptococcus sanguis*.

Sublist F

1. *Streptococcus sanguis*
2. *Streptococcus pneumoniae*
3. *Streptococcus faecalis*
4. *Streptococcus pyogenes*
5. *Streptococcus mutans*

Appendix B.   Classification of Microorganisms on the Basis of Hazard

Appendix B-I.   Classification of Etiologic Agents.   The original reference for this classification was the publication *Classification of Etiological Agents on the Basis of Hazard,* 4th edition, July 1974, U.S. Department of Health, Education, and Welfare, Public Health Service, Center for Disease Control, Office of Biosafety, Atlanta, Georgia 30333. For the purposes of these Guidelines, this list has been revised by the NIH (1).

Appendix B-I-A.   Class 1 Agents.   All bacterial, parasitic, fungal, viral, rickettsial, and chlamydial agents not included in higher classes.

Appendix B-I-B.   Class 2 Agents.

Appendix B-I-B-1. Bacterial Agents.

Acinetobacter calcoaceticus
Actinobacillus—all species
Aeromonas hydrophila
Arizona hinshawii—all serotypes
Bacillus anthracis
Bordetella—all species
Borrelia recurrentis, B. vincenti
Campylobacter fetus
Campylobacter jejuni
Chlamydia psittaci
Chlamydia trachomatis
Clostridium botulinum, Cl. chauvoei, Cl. haemolyticum, Cl. histolyticum, Cl.
   novyi, Cl. septicum, Cl. tetani
Corynebacterium diphtheriae, C. equi, C. haemolyticum, C. pseudotuberculosis,
   C. pyogenes, C. renale
Edwardsiella tarda
Erysipelothrix insidiosa
Escherichia coli—all enteropathogenic, enterotoxigenic, enteroinvasive and strains
   bearing Kl antigen
Haemophilus ducreyi, H. influenzae
Klebsiello—all species and all serotypes
Legionella pneumophila
Leptospira intarrogans—all serotypes
Listerio—all species
Moraxelia—all species
Mycobacteria—all species except those listed in Class 3
Mycoplasma—all species except *Mycoplasma mycoides* and *Mycoplasma agalactiae,*
   which are in Class 5
Neisseria gonorrhoeae, N. meningitidis
Pasteurella—all species except those listed in Class 3
Salmonella—all species and all serotypes
Shigella—all species and all serotypes
Sphaerophorus necrophorus
Staphylococcus aureus
Streptobacillus moniliformis
Streptococcus pneumoniae
Streptococcus pyogenes
Treponema carcteum, T. pallidum, and *T. pertenue*
Vibrio cholerae
Vibrio parahemolyticus
Yersinia enterocolitica

Appendix B-I-B-2., Fungal Agents.

Actinomycetes (including *Nocardia* species, *Actinomyces* species, and *Arachnia pro-*
   pionica) (2)
Blastomyces dermatitidis
Cryptococcus neoformans
Paracoccidioides braziliensis

Appendix B-I-B-3. Parasitic Agents.

Endamoeba histolytica
Leishmania sp.
Naegleria gruberi
Schistosoma mansoni
Toxoplasma gondii

Toxocara canis
Trichinella spiralis
Trypanosoma cruzi

Appendix B-I-B-4.   Viral, Rickettsial, and Chlamydial Agents.

Adenoviruses—human—all types
Cache Valley virus
Coxsackie A and *B* viruses
Cytomegaloviruses
Echoviruses—all types
Encephalomyocarditis virus (EMC)
Flanders virus
Hart Park virus
Hepatitis—associated antigen material
Herpes viruses—except *Herpesvirus simiae (Monkey B virus)* which is in Class 4
Corona viruses
Influenza viruses—all types except A/PR8/34, which is in Class 1
Langat virus
Lymphogranuloma venereum agent
Measles virus
Mumps virus
Parainfluenza virus—all types except *Parainfluenza virus 3,* SF4 strain, which is in
    Class 1
Polioviruses—all types, wild and attenuated
Poxviruses—all types except *Alestrim, Smallpox,* and *Whitepox* which are Class 5 and
    Monkey pox which depending on experiments is in Class 3 or Class 4
Rabies virus—all strains except *Rabies street virus* which should be classified in Class 3
Reoviruses—all types
Respiratory syncytial virus
Rhinoviruses—all types
Rubella virus
Simian viruses—all types except *Herpesvirus simiae (Monkey B virus)* and *Marburg virus*
    which are in Class 4
Sindbis virus
Tensaw virus
Turlock virus
Vaccinia virus
Varicella virus
Vesicular stomatitis virus (3)
Vole rickettsia
Yellow fever virus, 17D vaccine strain

Appendix B-I-C.   Class 3 Agents

Appendix B-I-C-1.   Bacterial agents

Baronello—all species
Brucella—all species
Francisella tularensis
Mycobacterium avium. M. bovis. M. tuberculosis
Pasteurella multocide type B ("buffalo" and other foreign virulent strains) (3)
Pseudomonas mallei (3)
Pseudomonas pseudomallei (3)
Yersinia pestis

Appendix B-I-C-2. Fungal Agents

Coccidioides immitis
Histoplasma capsulatum
Histoplasma capsulatum var. *duboisii*

Appendix B-I-C-3.   Parasitic Agents

None.

Appendix B-I-C-4.   Viral, Rickettsial, and Chlamydia Agents

Monkey pox, when used *in vitro* (4)
Arboviruses—all strains except those in Class 2 and 4 (*Arboviruses* indigenous to the
    United States are in Class 3 except those listed in Class 2 *West Nile* and *Semliki Forest*
    viruses may be classified up or down depending on the conditions of use and
    geographical location of the laboratory).
Dengue virus, when used for transmission or animal inoculation experiments
Lymphocytic choriomeningitis virus (LCM)
Rabies street virus
Rickettsia—all species except *Vole rickettsia* when used for transmission nor animal
    inoculation experiments
Yellow fever virus—wild, when used *in vitro*

Appendix B-I-D.   Class 4 Agents

Appendix B-I-D-1.   Bacterial Agents

None.

Appendix B-I-D-2.   Fungal Agents

None.

Appendix B-I-D-3.   Parasitic Agents

None.

Appendix B-I-D-4.   Viral, Rickettsial, and Chlamydial Agents

Ebola fever virus
Monkey pox, when used for transmission or animal inoculation experiments (4)
Hemorrhagic fever agents, including *Crimeac hemorrhagic fever, (Congo), Junin,* and
    Machupo viruses, and others as yet undefined
Herpesvirus simiae (Monkey B virus)
Lassa virus
Markburg virus
Tick-borne encephalitis virus complex, including *Russian spring-summer encephalitis,*
    Kyasanur forest disease. Omsk hemorrhogic fever, and *Central European encephalitis*
    viruses
Venezuelan equine encephalitis virus, epidemic strains, when used for transmission or
    animal inoculation experiments
Yellow fever virus—wild, when used for transmission or animal inoculation experiments

Appendix B-II.   Classification of Oncogenic Viruses on the Basis of Potential
        Hazard (5)

Appendix B-II-A.   Low-Risk Oncogenic Viruses

Rous sarcoma
SV-40
CELO
AD7-SV40
Polyoma
Bovine papilloma
Rat mammary tumor

Avian leukosis
Murine sarcoma
Mouse mammary tumor
Rat leukemia
Hamster Leukemia
Bovine Leukemia
Dog sarcoma
Mason-Pfizer monkey virus
Marek's
Guinea pig herpes
Lucke (Frog)
Adenovirus
Shope fibroma
Shope papilloma

Appendix B-II-B. Moderate-Risk Oncogenic Viruses

Ad2-SV40
FeLV
HV Saimiri
EBV
SSV-1
GaLV
HV ateles
Yaba
FeSV

Appendix B-III. Class 5 Agents

Appendix B-III-A. Animal Disease Organisms Which are Forbidden Entry into the United States by Law

Foot and mouth disease virus.

Appendix B-III-B. Animal Disease Organisms and Vectors Which are Forbidden Entry into the United States by USDA Policy

African horse sickness virus
African swine fever virus
Besnoitia besnoiti
Borna disease virus
Bovine infectious petechial fever
Camel pox virus
Ephemeral fever virus
Fowl plague virus
Goat pox virus
Hog cholera virus
Louping ill virus
Lumpy skin disease virus
Nairobi sheep disease virus
Newcastle disease virus (Asiatic strains)
Mycoplasma (contagious bovine pleuropneumona)
Mycoplasma agalactiae (contagious agalactia of sheep)
Rickettsia ruminatium (heart water)
Rift valley fever virus
Rhinderpest virus
Sheep pox virus
Swine vesicular disease virus

Teschen disease virus
Trypanosoma vivax (Nagana)
Trypanosoma evansi
Theileria parva (East Coast fever)
Theileria annulata
Theileria lawrencei
Theileria bovis
Theileria hirci
Vesicular exanthema virus
Wesselsbron disease virus
Zyonema

Appendix B-III-C.   Organisms Which may not be Studied in the United States Except at Specified Facilities

Small pox (4)
Alastrim (4)
White pox (4)

Appendix B-IV.   Footnotes and References of Appendix B

(1)   The original reference for this classification was the publication *Classification of Etiologic Agents on the Basis of Hazard,* 4th edition, July 1974, U.S. Department of Health, Education, and Welfare, Public Health Service, Centers for Disease Control, Office of Biosafety, Atlanta, Georgia 30333. For the purposes of these Guidelines, this list has been revised by the NIH.

(2)   Since the publication of the classification in 1974 (1), the *Actinomycetes* have been reclassified as bacterial rather than fungal agents.

(3)   A USDA permit, required for import and interstate transport of pathogens, may be obtained from the Animal and Plant Health Inspection Service, USDA, Federal Building, Hyattsville, MD 20782.

(4)   All activities, including storage of variola and whitepox, are restricted to the single national facility (World Health Organization (WHO) Collaborating Center for Smallpox Research, Centers for Disease Control, in Atlanta).

(5)   *National Cancer Institute Safety Standards for Research Involving Oncogenic Viruses* (October 1974). U.S. Department of Health, Education, and Welfare Publication No. (NIH) 75-790.

(6)   U.S. Department of Agriculture, Animal, and Plant Health Inspection Service.

Appendix C.   Exemptions Under Section III-D-5

Section III-D-5 states that exempt from these Guidelines are "Other classes of recombinant DNA molecules if the Director, NIH, with advice of the RAC, after appropriate notice and opportunity for public comment finds that they do not present a significant risk to health or the environment (see Section IV-C-1-b-(1)-(c)). Certain classes are exempt as of publication of these revised Guidelines."

The following classes of experiments are exempt under Section III-D-5 of the Guidelines:

Appendix C-I.   Recombinant DNAs in Tissue Culture.   Recombinant DNA molecules derived entirely from non-viral components (that is, no component is derived from a eukaryotic virus) that are propagated and maintained in cells in tissue culture are exempt from these Guidelines with the exceptions listed below.

Exceptions

Experiments described in Section III-A which require specific RAC review and NIH approval before initiation of the experiment.

Experiments involving DNA from Class 3, 4, or 5 organisms (1) or cells known to be infected with these agents.

Experiments involving the deliberate introduction of genes coding for the biosynthesis of molecules toxic for vertebrates (see Appendix F).

Appendix C-II. Experiments Involving E. coli *K-12 Host-Vector Systems.* Experiments which use *E. coli* K-12 host-vector systems, with the exception of those experiments listed below, are exempt from these Guidelines provided that: (i) the *E. coli* host shall not contain conjugation proficient plasmids or generalized transducing phages, and (ii) lambda or lambdoid or Ff bacteriophages or nonconjugative plasmids (2) shall be used as vectors. However, experiments involving the insertion into *E. coli* K-12 of DNA from prokaryotes that exchange genetic information (3) with *E. coli* may be performed with any *E. coli* K-12 vector (e.g., conjugative plasmid). When a nonconjugative vector is used, the *E. coli* K-12 host may contain conjugation-proficient plasmids either autonomous or integrated, or generalized transducing phages.

For these exempt laboratory experiments, Bl1 physical containment conditions are recommended.

For large-scale (LS) fermentation experiments BL1-LS physical containment conditions are recommended. However, following review by the IBC of appropriate data for a particular host-vector system, some latitude in the application of BL1-LS requirements as outlined in Appendix K-II-A through K-II-F is permitted.

Exceptions

Experiments described in Section III-A which require specific RAC review and NIH approval before initiation of the experiment.

Experiments involving DNA from Class 3, 4, or 5 organisms (1) or from cells known to be infected with these agents may be conducted under containment conditions specified in Section III-B-2 with prior IBC review and approval.

Large-scale experiments (e.g., more than 10 liters of culture) require prior IBC review and approval (see Section III-B-5).

Experiments involving the deliberate cloning of genes coding for the biosynthesis of molecules toxic for vertebrates (see Appendix F).

Appendix C-III. Experiments Involving Saccharomyces cerevisiae *Host Vector Systems.* Experiments which use *Saccharomyces cerevisiae* host-vector systems, with the excemption of experiments listed below, are exempt from these Guidelines provided that laboratory strains are used.

For these exempt laboratory experiments, BL1 physical containment conditions are recommended.

For large-scale fermentation experiments BL1-LS physical containment conditions are recommended. However, following review by the IBC or appropriate data for a particular host-vector system some latitude in the application of BL1-LS requirements as outlined in Appendix K-II-A through K-II-F is permitted.

Exceptions

Experiments described in Section III-A which require specific RAC review and NIH approval before initiation of the experiment.

Experiments involving Class 3, 4, or 5 organisms (1) or cells known to be infected with these agents may be conducted under containment conditions specified in Section III-B-2 with prior IBC review and approval

Large-scale experiments (e.g., more than 10 liters of culture) require prior IBC review and approval (see Section III-B-5).

Experiments involving the deliberate cloning of genes coding for the biosynthesis of molecules toxic for vertebrates (see Appendix F).

Appendix C-IV. Experiments Involving Bacillus subtilis *Host-Vector Systems.* Any asporogenic *Bacillus subtilis* strain which does not revert to a sporeformer with a frequency greater than 10^{-7} can be used for cloning DNA with the exception of these experiments listed below. Indigenous *Bacillus* plasmids and phages whose host-range include *Bacillus cereus* or *Bacillus anthracis* may be used as vectors.

For these exempt laboratory experiments, BL1 physical containment conditions are recommended.

For large-scale fermentation experiments BL1-LS physical containment conditions are recommended. However, following review by the IBC of appropriate data for a particular host-vector system, some latitude in the application of BL1-LS requirements as outlined in Appendix K-II-A through K-II-F is permitted.

Exceptions

Experiments described in Section III-A which require specific RAC review and approval before initiation of the experiment.

Experiments involving Class 3, 4, or 5 organisms (1) or cells known to be infected with these agents may be conducted under containment conditions specified by Section III-B-2 with prior IBC review and approval.

Large-scale experiments (e.g., more than 10 liters of culture) require prior IBC review and approval (see Section III-B-5).

Experiments involving the deliberate cloning of genes coding for the biosynthesis of molecules toxic for vertebrates (see Appendix F).

Appendix C-V. Footnotes and References of Appendix C

(1)   The original reference to organisms as Class 1, 2, 3, 4, or 5 refers to the classification in the publication *Classification of Etiologic Agents on the Basis of Hazard,* 4th edition, July 1974; U.S. Department of Health, Education, and Welfare, Public Health Service, Centers for Disease Control, Office of Biosafety, Atlanta, Georgia 30333.

The Director, NIH, with advice of the Recombinant DNA Advisory Committee, may revise the classification for the purposes of these Guidelines (see Section IV-C-1-b-(2)-(d)). The revised list of organisms in each class is reprinted in Appendix B to these guidelines.

(2)   A subset of non-conjugative plasmid vectors are also poorly mobilizable (e.g., pBR322, pBR313). Where practical, these vectors should be employed.

(3)   Defined as observable under optimal laboratory conditions by transformation, transduction, phage infection, and/or conjugation with transfer of phage, plasmid, and/or chromosomal genetic information. Note that this definition of exchange may be less stringent than that applied to exempt organisms under Section III-D-4.

Appendix D.   Actions Taken Under the Guidelines

As noted in the subsections of Section IV-C-1-b-(1), the Director, NIH, may take certain actions with regard to the Guidelines after the issues have been considered by the RAC. Some of the actions taken to date include the following:

Appendix D-I.   Permission is granted to clone foot and mouth disease virus in the EK1 host-vector system consisting of *E. coli* K-12 and the vector pBR322, all work to be done at the Plum Island Animal Disease Center.

Appendix D-II.   Certain specified clones derived from segments of the foot and mouth disease virus may be transferred from Plum Island Animal Disease Center to the facilities of Genentech, Inc., of South San Francisco, California. Further development of the clones at Genentech has been approved under BL1 + EK1 conditions.

Appendix D-III.   The Rd strain of *Haemophilus influenzae* can be used as a host for the propagation of the cloned Tn 10 tet R gene derived from *E. coli* K-12 employing the non-conjugative *Haemophilus* plasmid, PRSFO885, under BL1 conditions.

Appendix D-IV.   Permission is granted to clone certain subgenamic segments of foot and mouth disease virus in HV1 *Bacillus subtilis* and *Saccharomyces cerevisiae* host-vector systems under BL1 conditions at Genentech, Inc., South San Francisco, California.

Appendix D-V.   Permission is granted to Dr. Ronald Davis of Stanford University to field test corn plants modified by recombinant DNA techniques under specified containment conditions.

Appendix D-VI.   Permission is granted to clone in *E. coli* K-12 under BL1 physical containment conditions subgenomic segments of rift valley fever virus subject to conditions which have been set forth by the RAC.

Appendix D-VII.   Attenuated laboratory strains of *Salmonella typhimurium* may be used under BL1 physical containment conditions to screen for the *Saccharomyces cerevisiae* pseudouridine synthetase gene. The plasmid YEp13 will be employed as the vector.

Appendix D-VIII.   Permission is granted to transfer certain clones of subgenomic segments of foot and mouth disease virus from Plum Island Animal Disease Center to the laboratories of Molecular Genetics, Inc., Minnetonka, Minnesota, and to work with these clones under BL1 containment conditions. Approval is contingent upon review of data on infectivity testing of the clones by a working group of the RAC.

Appendix D-IX.   Permission is granted to Dr. John Sanford of Cornell University to field test tomato and tobacco plants transformed with bacterial (*E. coli* K-12) and yeast DNA using pollen as a vector.

Appendix D-X.   Permission is granted to Drs. Steven Lindow and Nickolas Panopoulos of the University of California, Berkeley, to release under specified conditions *Pseudomonas syringae* pv. *syringae* and *Erwinia herbicola* carrying *in vitro* generated deletions of all or part of the genes involved in ice nucleation.

Appendix E.   Certified Host-Vector Systems

(See also Appendix I.)

While many experiments using *E. coli* K-12, *Saccharomyces cerevisiae* and *Bacillus subtilis* are currently exempt from the Guidelines under Section III-D-5, some derivatives of these host-vector systems were previously classified as HV1 or HV2. A listing of those systems follows:

Appendix E-I.   Bacillus subtilis

HV1.   The following plasmids are accepted as the vector components of certified *B. subtilis* HV1 systems: pUB110, pC194, pS194, pSA2100, pE194, PT127, pUB112, pC221, pC223, and pAB124. *B. subtilis* strains RUB 331 and BGSC 1S53 have been certified as the host component of HV1 systems based on these plasmids.

HV2d. The asporogenic mutant derivative of *Bacillus subtilis,* ASB 298, with the following plasmids as the vector component: pUB110, pC194, pS194, pSA2100, pE194, pT127, pUB112, pC221, pC223, and pAB124.

Appendix E-II. Saccharomyces cerevisiae

HV2. The following sterile strains of *Saccharomyces cerevisiae,* all of which have the ste- VC9 mutation, SHY1, SHY2, SHY3, and SHY4. The following plasmids are certified for use: YIp1, YEp2, YEp4, YIp5, YEp6, YRp7, Yep20, YEp21, YEp24, YIp25, YIp26, YIp27, YIp28, YIp29, YIp30, YIp31, YIp32, and YIp33.

Appendix E-III. Escherichia coli

EK2 Plasmid Systems. The *E. coli* K-12 strain chi-1776. The following plasmids are certified for use: pSC101, pMB9, pBR313, pBR322, pDH24, pBR325, pBR327, pGL101, pHB1. The following *E. coli/S. cerevisiae* hybrid plasmids are certified as EK2 vectors when used in *E. coli* chi-1776 or in the sterile yeast strains, SHY1, SHY2, SHY3 and SHY4: YIp1, YEp2, YEp4, YIp5, YEp6, YRp7, YEp20, YEp21, YEp24, YIp25, YIp26, YIp27, YIp28, YIp29, YIp30, YIp31, YIp32, YIp33.

EK2 Bacteriophage Systems. The following are certified EK2 systems based on bacteriophage lambda:

Vector	*Host*
λgtWESλB′	DP50*sup*F.
λgtWESλB*	DP50*sup*F.
λgtZ1*vir*λB′	*E. coli* K-12.
λgtALOλB.	DP50*sup*F.
Charon 3A	DP50 or DP50*sup*F.
Charon 4A	DP50 or DP50*sup*F.
Charon 16A	DP50 or DP50*sup*F.
Charon 21A	DP50*sup*F.
Charon 23A	DP50 or DP50*sup*F.
Charon 24A	DP50 or DP50*sup*F.

E. coli K-12 strains chi-2447 and chi-2281 are certified for use with lambda vectors that are certified for use with strain DP50 or DP50*sup*F provided that the su⁻ strain will not be used as a propagation host.

Appendix E-IV. Neurospora crassa

HV1. The following specified strains of *Neurospora crassa* which have been modified to prevent aerial dispersion: Inl (inositolless) strains 37102, 37401, 46316, 64001, and 89601.

Csp-1 strain UCLA37 and csp-2 strains FS590, UCLA101 (these are conidial separation mutants).

Eas strain UCLA191 (an "easily wettable" mutant).

Appendix E-V. Streptomyces

HV1. The following *Streptomyces* species: *Streptomyces coelicolor, S. lividans, S. parvulus,* and *S. griseus.* The following are accepted as vector components of certified *Streptomyces* HV1 systems: *Streptomyces* plasmids SCP2, SLP1.2, pIJ101, actinophage phi C31, and their derivatives.

Appendix E-VI. Pseudomonas putida

HV1. *Pseudomonas putida* strain KT2440 with plasmid vectors pKT262, pKT263, and pKT264.

Appendix F. Containment Conditions for Cloning of Genes Coding for the Biosynthesis of Molecules Toxic for Vertebrates

Appendix F-I. General Information. Appendix F specifies the containment to be used for the deliberate cloning of genes coding for the biosynthesis of molecules toxic for vertebrates. The cloning of genes coding for molecules toxic for vertebrates that have an LD_{50} of less than 100 nanograms per kilogram body weight (e.g., microbial toxins such as the botulinum toxins, tetanus toxin, diphtheria toxin, *Shigella dysenteriae* neurotoxin) is covered under Section III-A-1 of the Guidelines and requires RAC review and NIH and IBC approval before initiation. No specific restrictions shall apply to the cloning of genes if the protein specified by the gene has an LD_{50} of 100 micrograms or more per kilogram of body weight. Experiments involving genes coding for toxic molecules with an LD_{50} of 100 micrograms or less per kilogram body weight shall be registered with ORDA prior to initiating the experiments. A list of toxic molecules classified as to LD_{50} is available from ORDA. Testing procedures for determining toxicity of toxic molecules not on the list are available from ORDA. The results of such tests shall be forwarded to ORDA which will consult with the RAC Working Group on Toxins prior to inclusion of the molecules on the list (see Section IV-C-1-b-(2)-(e)).

Appendix F-II. Containment Conditions for Cloning of Toxic Molecule Genes in E. coli *K-12*

Appendix F-II-A. Cloning of genes coding for molecules toxic for vertebrates that have an LD_{50} in the range of 100 nanograms to 1000 nanograms per kilogram body weight (e.g., abrin, *Clostridium perfringens* epsilon toxin) may proceed under BL2 + EK2 or BL3 + EK1 containment conditions.

Appendix F-II-B. Cloning of genes for the biosynthesis of molecules toxic for vertebrates that have an LD_{50} in the range of 100 micrograms per kilogram body weight may proceed under BL1 + EK1 containment conditions (e.g., *Staphylococcus aureus* alpha toxin, *Staphylococcus aureus* beta toxin, ricin, *Pseudomonas aeruginosa* exotoxin A, *Bordatella pertussis* toxin, the lethal factor of *Bacillus anthracis,* the *Pasteurella pestis* murine toxins, the oxygen-labile hemolysins such as streptolysin O, and certain neurotoxins present in snake venoms and other venoms).

Appendix F-II-C. Some enterotoxins are substantially more toxic when administered enterally than parenterally. The following enterotoxins shall be subject to BL1 + Ek1 containment conditions: cholera toxin, the heat labile toxins of *E. coli, Klebsiella,* and other related proteins that may be identified by neutralization with an antiserum monospecific for cholera toxin, and the heat stable toxins of *E. coli* and of *Yersinia enterocolitica.*

Appendix F-III. Containment Conditions for Cloning of Toxic Molecule Genes in Organisms Others than E. coli *K-12.* Requests involving the cloning of genes coding for molecules toxic for vertebrates in host-vector systems other than *E. coli* K-12 will be evaluated by ORDA which will consult with the Working Group on Toxins (see Section IV-C-1-b-(3)-(f)).

Appendix F-IV. Specific Approvals

Appendix F-IV-A. Permission is granted to clone the Exotoxin A gene of *Pseudomonas aeruginosa* under BL1 conditions in *Pseudomonas aeruginosa* and in *Pseudomonas putida.*

Appendix F-IV-B. The pyrogenic exotoxin type A (Tox A) gene of *Staphylococcus aureus* may be cloned in an HV2 *Bacillus subtilis* host-vector system under BL3 containment conditions.

Appendix F-IV-C. Restriction fragments of *Corynephage Beta* carrying the structural gene for diphtheria toxin may be safely cloned in *E. coli* K-12 in high containment Building 550 at the Frederick Cancer Research Facility. Laboratory practices and containment equipment are to be specified by the IBC. If the investigators wish to proceed with the experiments, a prior review will be conducted to advise NIH whether the proposal has sufficient scientific merit to justify the use of the NIH BL4 facility.

Appendix F-IV-D. The genes coding for the *Staphylococcus aureus* determinants, A, B, and F, which may be implicated in toxic shock syndrome may be cloned in *E. coli* K-12 under BL2 + EK1 conditions. The *Staphylococcus aureus* strain used as the donor is to be alpha toxin minus. It is suggested that, if possible, the donor *Staphylococcus aureus* strain should lack other toxins with LD_{50}s in the range of one microgram per kilogram body weight such as the exfoliative toxin.

Appendix F-IV-E. Fragments F-1, E-2, and F-3 of the diphtheria toxin gene *(tox)* may be cloned in *E. coli* K-12 under BL1 + EK1 containment conditions and may be cloned in *Bacillus subtilis* host-vector systems under BL1 containment conditions. Fragment F-1 and fragment F-2 both contain: (i) some or all of the transcriptional control elements of *tox;* (ii) the signal peptide; and (iii) fragment A (the center responsible for ADP-ribosylation of elongation factor 2). Fragment F-3 codes for most of the non-toxic fragment B of the toxin and contains no sequences coding for any portion of the enzymatically-active fragment A moiety.

Appendix F-IV-F. The gene(s) coding for a toxin (designated LT-like) isolated from *E. coli* which is similar to the *E. coli* heat labile enterotoxin (LT) with respect to its activities and mode of action but is not neutralized by antibodies against cholera enterotoxin or against LT from human or porcine *E. coli* strains, and sequences homologous to the *E. coli* LT-like toxin gene may be cloned under BL1 + EK1 conditions.

Appendix F-IV-G. Genes from *Vibrio fluvialis, Vibrio mimicus,* and non *0-1 Vibrio cholerae,* specifying virulence factors for animals, may be cloned under BL1 + EK1 conditions. The virulence factors to be cloned will be selected by testing fluid induction in suckling mice and Y-1 mouse adrenal cells.

Appendix F-IV-H. The intact structural gene(s) of the Shiga-like toxin from *E. coli* may be cloned in *E. coli* K-12 under BL3 + EK1 containment conditions.

E. coli host-vector systems expressing the Shiga-like toxin gene product may be moved from BL3 + BL2 containment conditions provided that: (1) the amount of toxin produced by the modified host-vector systems be no greater than that produced by the positive control strain 933 *E. coli* 0157H7, grown and measured under optimal conditions; and (2) the cloning vehicle is to be an EK1 vector preferably belonging to the class of poorly mobilizable plasmids such as pBR322, pBR328, and pBR325.

Nontoxinogenic fragments of the Shiga-like toxin structural gene(s) may be moved from BL3 + EK1 to BL2 + EK1 containment conditions or such nontoxic fragments may be directly cloned in *E. coli* K-12 under BL2 + EK1 conditions provided that the *E. coli* host-vector systems containing the fragments do not contain overlapping fragments which together would encompass the Shiga-like toxin structural gene(s).

Appendix F-IV-I. A hybrid gene in which the gene coding for the melanocyte stimulating hormone (MSH) is joined to a segment of the gene encoding diphtheria toxin may be safely propagated in *E. coli* K-12 under BL4 containment in high containment building 550 at the Frederick Cancer Research Facility. If the investigators wish to proceed with the experiment, a prior review will be conducted to advise NIH whether the proposal has sufficient scientific merit to justify the use of the NIH BL4 facility. Before any of the strains may be removed from the BL4 facility, data on their safety shall be evaluated by the Working Group on Toxins and the working group recommendation shall be acted upon by NIH.

Appendix F-IV-J. The gene segment encoding the A subunit of cholera toxin of *Vibrio cholerae* may be joined to the transposons Tn5 and Tn5-131 and the A-subunit: Tn5-131 hybrid gene cloned in *E. coli* K-12 and *V. cholerae* under BL1 containment conditions.

Appendix G. Physical Containment

Appendix G-I. Standard Practices and Training. The first principle of containment is a strict adherence to good microbiological practices (1-10). Consequently, all personnel directly or indirectly involved in experiments on recombinant DNAs must receive adequate instruction (see Sections IV-B-1-e and IV-B-5-d). This shall, as a minimum, include instructions in aseptic techniques and in the biology of the organisms used in the experiments so that the potential biohazards can be understood and appreciated.

Any research group working with agents with a known or potential biohazard shall have an emergency plan which describes the procedures to be followed if an accident contaminates personnel or the environment. The PI must ensure that everyone in the laboratory is familiar with both the potential hazards of the work and the emergency plan (see Sections IV-B-3-d and IV-B-5-e). If a research group is working with a known pathogen for which there is an effective vaccine, the vaccine should be made available to all workers. Where serological monitoring is clearly appropriate, it shall be provided (see Section IV-B-1-f).

The "Laboratory Safety Monograph" and *Biosafety in Microbiological and Biomedical Laboratories (2)* booklets describe practices, equipment, and facilities in detail.

Appendix G-II. Physical Containment Levels. The objective of physical containment is to confine organisms containing recombinant DNA molecules and thus to reduce the potential for exposure of the laboratory worker, persons outside of the laboratory, and the environment to organisms containing recombinant DNA molecules. Physical containment is achieved through the use of laboratory practices, containment equipment, and special laboratory design. Emphasis is placed on primary means of physical containment which are provided by laboratory practices and containment equipment. Special laboratory design provides a secondary means of protection against the accidental release of organisms outside the laboratory or to the environment. Special laboratory design is used primarily in facilities in which experiments of moderate to high potential hazards are performed.

Combinations of laboratory practices, containment equipment, and special laboratory design can be made to achieve different levels of physical containment. Four levels of physical containment, which are designated as BL1, BL2, BL3, and BL4, are described. It should be emphasized that the descriptions and assignments of physical containment detailed below are based on existing approaches to containment of pathogenic organisms (2). The National Cancer Institute describes three levels for research on oncogenic viruses which roughly correspond to our BL2, BL3, and BL4 levels (3).

It is recognized that several different combinations of laboratory practices, containment equipment, and special laboratory design may be appropriate for containment of specific research activities. The Guidelines, therefore, allow alternative selections of primary containment equipment within facilities that have been designed to provide BL3 and BL4 levels of physical containment. The selection of alternative methods of primary containment is dependent, however, on the level of biological containment provided by the host-vector system used in the experiment. Consideration will also be given by the Director, NIH, with the advice of the RAC to other combinations which achieve an equivalent level of containment (see Section IV-C-1-b-(2)-(b)).

Appendix G-II-A.   Biosafety Level 1 (BL1) (13)

Appendix G-II-A-1.   Standard Microbiological Practices

Appendix G-II-A-1-a.   Access to the laboratory is limited or restricted at the discretion of the laboratory director when experiments are in progress.

Appendix G-II-A-1-b.   Work surfaces are decontaminated once a day and after any spill of viable material.

Appendix G-II-A-1-c.   All contaminated liquid or solid wastes are decontaminated before disposal.

Appendix G-II-A-1-d.   Mechanical-pipetting devices are used; mouth pipetting is prohibited.

Appendix G-II-A-1-e.   Eating, drinking, smoking, and applying cosmetics are not permitted in the work area. Food may be stored in cabinets or refrigerators designated and used for this purpose only.

Appendix G-II-A-1-f.   Persons wash their hands after they handle materials involving organisms containing recombinant DNA molecules, and animals, and before leaving the laboratory.

Appendix G-II-A-1-g.   All procedures are performed carefully to minimize the creation of aerosols.

Appendix G-II-A-1-h.   It is recommended that laboratory coats, gowns, or uniforms be worn to prevent contamination or soiling of street clothes.

Appendix G-II-A-2.   Special Practices

Appendix G-II-A-2-a.   Contaminated materials that are to be decontaminated at a site away from the laboratory are placed in a durable leakproof container which is closed before being removed from the laboratory.

Appendix G-II-A-2-b.   An insect and rodent control program is in effect.

Appendix G-II-A-3.   Containment Equipment.

Appendix G-II-A-3-a.   Special containment equipment is generally not required for manipulations of agents assigned to Biosafety Level 1.

Appendix G-II-A-4.   Laboratory Facilities.

Appendix G-II-A-4-a.   The laboratory is designed so that it can be easily cleaned.

Appendix G-II-A-4-b.   Bench tops are impervious to water and resistant to acids, alkalis, organic solvents, and moderate heat.

Appendix G-II-A-4-c.   Laboratory furniture is sturdy. Spaces between benches, cabinets, and equipment are accessible for cleaning.

Appendix G-II-A-4-d.   Each laboratory contains a sink for hand-washing.

Appendix G-II-A-4-e.   If the laboratory has windows that open, they are fitted with fly screens.

Appendix G-II-B.   Biosafety Level 2 (BL2) (14)

Appendix G-II-B-1.   Standard Microbiological Practices.

Appendix G-II-B-1-a.   Access to the laboratory is limited or restricted by the laboratory director when work with organisms containing recombinant DNA molecules is in progress.

Appendix G-II-B-1-b. Work surfaces are decontaminated at least once a day and after any spill of viable material.

Appendix G-II-B-1-c. All contaminated liquid or solid wastes are decontaminated before disposal.

Appendix G-II-B-1-d. Mechanical pipetting devices are used; mouth pipetting is prohibited.

Appendix G-II-B-1-e. Eating, drinking, smoking, and applying cosmetics are not permitted in the work area. Food may be stored in cabinets or refrigerators designated and used for this purpose only.

Appendix G-II-B-1-f. Persons wash their hands after handling materials involving organisms containing recombinant DNA molecules, and animals, and when leaving the laboratory.

Appendix G-II-B-1-g. All procedures are performed carefully to minimize the creation of aerosols.

Appendix G-II-B-1-h. Experiments of lesser biohazard potential can be carried out concurrently in carefully demarcated areas of the same laboratory.

Appendix G-II-B-2. Special Practices

Appendix G-II-B-2-a. Contaminated materials that are to be decontaminated at a site away from the laboratory are placed in a durable leakproof container which is closed before being removed from the laboratory.

Appendix G-II-B-2-b. The laboratory director limits access to the laboratory. The director has the final responsibility for assessing each circumstance and determining who may enter or work in the laboratory.

Appendix G-II-B-2-c. The laboratory director establishes policies and procedures whereby only persons who have been advised of the potential hazard and meet any specific entry requirements (e.g., immunization) enter the laboratory or animal rooms.

Appendix G-II-B-2-d. When the organisms containing recombinant DNA molecules in use in the laboratory require special provisions for entry (e.g., vaccination, a hazard warning sign incorporating the universal biohazard symbol is posted on the access door to the laboratory work area. The hazard warning sign identifies the agent, lists the name and telephone number of the laboratory director or other responsible person(s), and indicates the special requirement(s) for entering the laboratory.

Appendix G-II-B-2-e. An insect and rodent control program is in effect.

Appendix G-II-B-2-f. Laboratory coats, gowns, smocks, or uniforms are worn while in the laboratory. Before leaving the laboratory for nonlaboratory areas (e.g., cafeteria, library, administrative offices), this protective clothing is removed and left in the laboratory or covered with a clean coat not used in the laboratory.

Appendix G-II-B-2-g. Animals not involved in the work being performed are not permitted in the laboratory.

Appendix G-II-B-2-h. Special care is taken to avoid skin contamination with organisms containing recombinant DNA molecules; gloves should be worn when handling experimental animals and when skin contact with the agent is unavoidable.

Appendix G-II-B-2-i. All wastes from laboratories and animal rooms are appropriately decontaminated before disposal.

Appendix G-II-B-2-j. Hypodermic needles and syringes are used only for parenteral injection and aspiration of fluids from laboratory animals and diaphragm bottles. Only needle-locking syringes or disposable syringe-needle units (i.e., needle is integral to the syringe) are used for the injection or aspiration of fluids containing organisms that contain recombinant DNA molecules. Extreme caution should be used when handling needles and syringes to avoid autoinoculation and the generation of aerosols during use and disposal. Needles should not be bent, sheared, replaced in the needle sheath or guard or removed from the syringe following use. The needle and syringe should be promptly placed in a puncture-resistant container and decontaminated, preferably by autoclaving, before discard or reuse.

Appendix G-II-B-2-k. Spills and accidents which result in overt exposures to organisms containing recombinant DNA molecules are immediately reported to the laboratory director. Medical evaluation, surveillance, and treatment are provided as appropriate and written records are maintained.

Appendix G-II-B-2-l. When appropriate, considering the agent(s) handled, baseline serum samples for laboratory and other at-risk personnel are collected and stored. Additional serum specimens may be collected periodically depending on the agents handled or the function of the facility.

Appendix G-II-B-2-m. A biosafety manual is prepared or adopted. Personnel are advised of special hazards and are required to read instructions on practices and procedures and to follow them.

Appendix G-II-B-3. Containment Equipment

Appendix G-II-B-3-a. Biological safety cabinets (Class I or II) (see Appendix G-III-12) or other appropriate personal protective or physical containment devices are used whenever:

Appendix G-II-B-3-a-(1). Procedures with a high potential for creating aerosols are conducted (15). These may include centrifuging, grinding, blending, vigorous shaking or mixing, sonic disruption, opening containers of materials whose internal pressures may be different from ambient pressures, inoculating animals intranasally, and harvesting infected tissues from animals or eggs.

Appendix G-II-B-3-a-(2). High concentrations or large volumes of organisms containing recombinant DNA molecules are used. Such materials may be centrifuged in the open laboratory if sealed heads or centrifuge safety cups are used and if they are opened only in a biological safety cabinet.

Appendix G-II-B-4. Laboratory Facilities

Appendix G-II-B-4-a. The laboratory is designed so that it can be easily cleaned.

Appendix G-II-B-4-b. Bench tops are impervious to water and resistant to acids, alkalis, organic solvents, and moderate heat.

Appendix G-II-B-4-c. Laboratory furniture is sturdy and spaces between benches, cabinets, and equipment are accessible for cleaning.

Appendix G-II-B-4-d. Each laboratory contains a sink for hand-washing.

Appendix G-II-B-4-e. If the laboratory has windows that open, they are fitted with fly screens.

Appendix G-II-B-4-f. An autoclave for decontaminating laboratory wastes is available.

Appendix G-II-C. Biosafety Level 3 (BL3) (16)

Appendix G-II-C-1. Standard Microbiological Practices.

Appendix G-II-C-1-a. Work surfaces are decontaminated at least once a day and after any spill of viable material.

Appendix G-II-C-1-b. All contaminated liquid or solid wastes are decontaminated before disposal.

Appendix G-II-C-1-c. Mechanical pipetting devices are used; mouth pipetting is prohibited.

Appendix G-II-C-1-d. Eating, drinking, smoking, storing food, and applying cosmetics are not permitted in the work area.

Appendix G-II-C-1-e. Persons wash their hands after handling materials involving organisms containing recombinant DNA molecules, and animals, and when they leave the laboratory.

Appendix G-II-C-1-f. All procedures are performed carefully to minimize the creation of aerosols.

Appendix G-II-C-1-g. Persons under 16 years of age shall not enter the laboratory.

Appendix G-II-C-1-h. If experiments involving other organisms which require lower levels of containment are to be conducted in the same laboratory concurrently with experiments requiring BL3 level physical containment, they shall be conducted in accordance with all BL3 level laboratory practices.

Appendix G-II-C-2.   Special Practices.

Appendix G-II-C-2-a. Laboratory doors are kept closed when experiments are in progress.

Appendix G-II-C-2-b. Contaminated materials that are to be decontaminated at a site away from the laboratory are placed in a durable leakproof container which is closed before being removed from the laboratory.

Appendix G-II-C-2-c. The laboratory director controls access to the laboratory and restricts access to persons whose presence is required for program or support purposes. The director has the final responsibility for assessing each circumstance and determining who may enter or work in the laboratory.

Appendix G-II-C-2-d. The laboratory director establishes policies and procedures whereby only persons who have been advised of the potential biohazard, who meet any specific entry requirements (e.g., immunization), and who comply with all entry and exit procedures enter the laboratory or animal rooms.

Appendix G-II-C-2-e. When organisms containing recombinant DNA molecules or experimental animals are present in the laboratory or containment module, a hazard warning sign incorporating the universal biohazard symbol is posted on all laboratory and animal room access doors. The hazard warning sign identifies the agent, lists the name and telephone number of the laboratory director or other responsible person(s), and indicates any special requirements for entering the laboratory, such as the need for immunizations, respirators, or other personal protective measures.

Appendix G-II-C-2-f. All activities involving organisms containing recombinant DNA molecules are conducted in biological safety cabinets or other physical containment devices within the containment module. No work in open vessels is conducted on the open bench.

Appendix G-II-C-2-g. The work surfaces of biological safety cabinets and other containment equipment are decontaminated when work with organisms containing recombinant DNA molecules is finished. Plastic-backed paper towelling used on nonperforated work surfaces within biological safety cabinets facilitates clean-up.

Appendix G-II-C-2-h. An insect and rodent control program is in effect.

Appendix G-II-C-2-i. Laboratory clothing that protects street clothing (e.g., solid front and wrap-around gowns, scrub suits, coveralls) is worn in the laboratory. Laboratory clothing is not worn outside the laboratory, and it is decontaminated before being laundered.

Appendix G-II-C-2-j. Special care is taken to avoid skin contamination with contaminated materials; gloves should be worn when handling infected animals and when skin contact with infectious materials is unavoidable.

Appendix G-II-C-2-k. Molded surgical masks or respirators are worn in rooms containing experimental animals.

Appendix G-II-C-2-l. Animals and plants not related to the work being conducted are not permitted in the laboratory.

Appendix G-II-C-2-m. Laboratory animals held in a BL3 area shall be housed in partial-containment caging systems, such as Horsfall units (11), open cages placed in ventilated enclosures, solid-wall and -bottom cages covered by filter bonnets, or solid-wall and -bottom cages placed on holding racks equipped with ultraviolet in radiation lamps and reflectors.

Note. Conventional caging systems may be used provided that all personnel wear appropriate personal protective devices. These shall include at a minimum wrap-around gowns, head covers, gloves, shoe covers, and respirators. All personnel shall shower on exit from areas where these devices are required.

Appendix G-II-C-2-n. All wastes from laboratories and animal rooms are appropriately decontaminated before disposal.

Appendix G-II-C-2-o. Vacuum lines are protected with high efficiency particulate air (HEPA) filters and liquid disinfectant traps.

Appendix G-II-C-2-p. Hypodermic needles and syringes are used only for parenteral injection and aspiration of fluids from laboratory animals and diaphragm bottles. Only needle-locking syringes and disposable syringe-needle units (i.e., needle is integral to the syringe) are used for the injection or aspiration of fluids containing organisms that contain recombinant DNA molecules. Extreme caution should be used when handling needles and syringes to avoid autoinoculation and the generation of aerosols during use and disposal. Needles should not be bent, sheared, replaced in the needle sheath or guard or removed from the syringe following use. The needle and syringe should be promptly placed in a puncture-resistant container and decontaminated, preferably by autoclaving, before discard or reuse.

Appendix G-II-C-2-q. Spills and accidents which result in overt or potential exposures to organisms containing recombinant DNA molecules are immediately reported to the laboratory director. Appropriate medical evaluation, surveillance, and treatment are provided and written records are maintained.

Appendix G-II-C-2-r. Baseline serum samples for all laboratory and other at-risk personnel should be collected and stored. Additional serum specimens may be collected periodically depending on the agents handled or the function of the laboratory.

Appendix G-II-C-2-s. A biosafety manual is prepared or adopted. Personnel are advised of special hazards and are required to read instructions on practices and procedures and to follow them.

Appendix G-II-C-2-t. Alternative Selection of Containment Equipment. Experimental procedures involving a host-vector system that provides a one-step higher level of biological containment than that specified can be conducted in the BL3 laboratory using containment equipment specified for the BL2 level of physical containment. Experimental procedures involving a host-vector system that provides a one-step lower level of biological containment than that specified can be conducted in the BL3 laboratory using containment equipment specified for the BL4 level of physical containment. Alternative combination of containment safeguards are shown in Table 1.

Appendix G-II-C-3. Containment Equipment

Appendix G-II-C-3-a. Biological safety cabinets (Class I, II, or III) (see Appendix G-III-12) or other appropriate combinations of personal protective or physical containment devices (e.g., special protective clothing, masks, gloves, respirators, centrifuge safety cups, sealed centrifuge rotors, and containment caging for animals) are used for all activities with organisms containing recombinant DNA molecules which pose a threat of aerosol exposure. These include: manipulation of cultures and of those clinical or environmental materials which may be a source of aerosols; the aerosol challenge of experimental animals; and harvesting infected tissues or fluids from experimental animals and embryonate eggs, and necropsy of experimental animals.

Appendix G-II-C-4.   Laboratory Facilities

Appendix G-II-C-4-a.   The laboratory is separated from areas which are open to unrestricted traffic flow within the building. Passage through two sets of doors is the basic requirement for entry into the laboratory from access corridors or other contiguous areas. Physical separation of the high containment laboratory from access corridors or other laboratories or activities may also be provided by a double-doored clothes change room (showers may be included), airlock, or other access facility which requires passage through two sets of doors before entering the laboratory.

Appendix G-II-C-4-b.   The interior surfaces of walls, floors, and ceilings are water resistant so that they can be easily cleaned. Penetrations in these surfaces are sealed or capable of being sealed to facilitate decontaminating the area.

Appendix G-II-C-4-c.   Bench tops are impervious to water and resistant to acids, alkalis, organic solvents, and moderate heat.

Appendix G-II-C-4-d.   Laboratory furniture is sturdy and spaces between benches, cabinets, and equipment are accessible for cleaning.

Appendix G-II-C-4-e.   Each laboratory contains a sink for hand-washing. The sink is foot, elbow, or automatically operated and is located near the laboratory exit door.

Appendix G-II-C-4-f.   Windows in the laboratory are closed and sealed.

Appendix G-II-C-4-g.   Access doors to the laboratory or containment module are self-closing.

Appendix G-II-C-4-h.   An autoclave for decontaminating laboratory wastes is available preferably within the laboratory.

Appendix G-II-C-4-i.   A ducted exhaust air ventilation system is provided. This system creates directional airflow that draws air into the laboratory through the entry area. The exhaust air is not recirculated to any other area of the building, is discharged to the outside, and is dispersed away from the occupied areas and air intakes. Personnel must verify that the direction of the airflow (into the laboratory) is proper. The exhaust air from the laboratory room can be discharged to the outside without being filtered or otherwise treated.

Appendix G-II-C-4-j.   The HEPA-filtered exhaust air from Class 1 or Class II biological safety cabinets is discharged directly to the outside or through the building exhaust system. Exhaust air from Class I or II biological safety cabinets may be circulated within the laboratory if the cabinet is tested and certified at least every twelve months. If the HEPA-filtered exhaust air from Class I or II biological safety cabinets is to be discharged to the outside through the building exhaust air system, it is connected to this system in a manner (e.g., thimble unit connection (12)) that avoids any interference with the air balance of the cabinets or building exhaust system.

Appendix G-II-D.   Biosafety Level 4 (BL4)

Appendix G-II-D-1.   Standard Microbiological Practices

Appendix G-II-D-1-a.   Work surfaces are decontaminated at least once a day and immediately after any spill of viable material.

Appendix G-II-D-1-b.   Only mechanical pipetting devices are used.

Appendix G-II-D-1-c.   Eating, drinking, smoking, storing food, and applying cosmetics are not permitted in the laboratory.

Appendix G-II-D-1-d. All procedures are performed carefully to minimize the creation of aerosols.

Appendix G-II-D-2. Special Practices

Appendix G-II-D-2-a. Biological materials to be removed from the Class III cabinets or from the maximum containment laboratory in a viable or intact state are transferred to a nonbreakable, sealed primary container and then enclosed in a nonbreakable, sealed secondary container which is removed from the facility through a disinfectant dunk tank, fumigation chamber, or an airlock designed for this purpose.

Appendix G-II-D-2-b. No materials, except for biological materials that are to remain in a viable or intact state, are removed from the maximum containment laboratory unless they have been autoclaved or decontaminated before they leave the facility. Equipment or material which might be damaged by high temperatures or steam is decontaminated by gaseous or vapor methods in an airlock or chamber designed for this purpose.

Appendix G-II-D-2-c. Only persons whose presence in the facility or individual laboratory rooms is required for program or support purposes are authorized to enter. The supervisor has the final responsibility for assessing each circumstance and determining who may enter or work in the laboratory. Access to the facility is limited by means of secure, locked doors; accessibility is managed by the laboratory director, biohazards control officer, or other person responsible for the physical security of the facility. Before entering, persons are advised of the potential biohazards and instructed as to appropriated safeguards for ensuring their safety. Authorized persons comply with the instructions and all other applicable entry and exit procedures. A logbook signed by all personnel indicates the date and time of each entry and exit. Practical and effective protocols for emergency situations are established.

Appendix G-II-D-2-d. Personnel enter and leave the facility only through the clothing change and shower rooms. Personnel shower each time they leave the facility. Personnel use the airlocks to enter or leave the laboratory only in an emergency.

Appendix G-II-D-2-e. Street clothing is removed in the outer clothing change room and kept there. Complete laboratory clothing, including undergarments, pants, and shirts or jumpsuits, shoes, and gloves, is provided and used by all personnel entering the facility. Head covers are provided for personnel who do not wash their hair during the exit shower. When leaving the laboratory and before proceeding into the shower area, personnel remove their laboratory clothing and store it in a locker or hamper in the inner change room.

Appendix G-II-D-2-f. When materials that contain organisms containing recombinant DNA molecules or experimental animals are present in the laboratory or animal rooms, a hazard warning sign incorporating the universal biohazard symbol is posted on all access doors. The sign identifies the agent, lists the name of the laboratory director or other responsible person(s), and indicates any special requirements for entering the area (e.g., the need for immunizations or respirators).

Appendix G-II-D-2-g. Supplies and materials needed in the facility are brought in by way of the double-doored autoclave, fumigation chamber, or airlock which is appropriately decontaminated between each use. After securing the outer doors, personnel within the facility retrieve the materials by opening the interior doors or the autoclave, fumigation chamber, or airlock. These doors are secured after materials are brought into the facility.

Appendix G-II-D-2-h. An insect and rodent control program is in effect.

Appendix G-II-D-2-i. Materials (e.g., plants, animals, and clothing) not related to the experiment being conducted are not permitted in the facility.

Appendix G-II-D-2-j. Hypodermic needles and syringes are used only for parenteral injection and aspiration of fluids from laboratory animals and diaphragm bottles. Only needle-locking syringes or disposable syringe-needle units (i.e., needle is integral part of unit) are used for the injection or aspiration of fluids containing organisms that contain recombinant DNA molecules. Needles should not be bent, sheared, replaced in the needle sheath or guard or removed from the syringe following use. The needle and syringe should be placed in a puncture-resistant container and decontaminated, preferably by autoclaving before discard or reuse. Whenever possible, cannulas are used instead of sharp needles (e.g., gavage).

Appendix G-II-D-2-k. A system is set up for reporting laboratory accidents and exposures and employee absenteeism and for the medical surveillance of potential laboratory-associated illnesses. Written records are prepared and maintained. An essential adjunct to such a reporting-surveillance system is the availability of a facility for quarantine, isolation, and medical care of personnel with potential or known laboratory associated illnesses.

Appendix G-II-D-2-l. Laboratory animals involved in experiments requiring BL4 level physical containment shall be housed either in cages contained in Class III cabinets or in partial containment caging systems (such as Horsfall units (11)), open cages placed in ventilated enclosures, or solid-wall and -bottom cages placed on holding racks equipped with ultraviolet irradiation lamps and reflectors that are located in a specially designed area in which all personnel are required to wear one-piece positive pressure suits.

Appendix G-II-D-2-m. Alternative Selection of Containment Equipment. Experimental procedures involving a host-vector system that provides a one-step higher level of biological containment than that specified can be conducted in the BL4 facility using containment equipment requirements specified for the BL3 level of physical containment. Alternative combinations of containment safeguards are shown in Table I.

Table 1. Possible alternate combinations of physical and biological containment safeguards

Classification of physical and biological containment	Alternate physical containment			Alternate biological containment
	Laboratory facilities	Laboratory practices	Containment equipment	
BL3/HV2	BL3	BL3	BL3	HV2
	BL3	BL3	BL4	HV1
BL3/HV1	BL3	BL3	BL3	HV1
	BL3	BL3	BL2	HV2
BL4/HV1	BL4	BL4	BL4	HV1
	BL4	BL4	BL3	HV2

Appendix G-II-D-3. Containment Equipment

Appendix G-II-D-3-a. All procedures within the facility with agents assigned to Biosafety Level 4 are conducted in the Class III biological safety cabinet or in Class I or II biological safety cabinets used in conjunction with one-piece positive pressure personnel suits ventilated by a life-support system.

Appendix G-II-D-4. Laboratory Facilities

Appendix G-II-D-4-a. The maximum containment facility consists of either a separate building or a clearly demarcated and isolated zone within a building. Outer and inner change rooms separated by a shower are provided for personnel entering and leaving the facility. A double-doored autoclave, fumigation chamber, or ventilated airlock is provided for passage of those materials, supplies, or equipment which are not brought into the facility through the change room.

Appendix G-II-D-4-b. Walls, floors, and ceilings of the facility are constructed to form a sealed internal shell which facilitate fumigation and is animal and insect proof. The internal surfaces of this shell are resistant to liquids and chemicals, thus facilitating cleaning and decontamination of the area. All penetrations in these structures and surfaces are sealed. Any drains in the floors contain traps filled with a chemical disinfectant of demonstrated efficacy against the target agent, and they are connected directly to the liquid waste decontamination system. Sewer and other ventilation lines contain HEPA filters.

Appendix G-II-D-4-c. Internal facility appurtenances, such as light fixtures, air ducts, and utility pipes, are arranged to minimize the horizontal surface area on which dust can settle.

Appendix G-II-D-4-d. Bench tops have seamless surfaces which are impervious to water and resistant to acids, alkalis, organic solvents, and moderate heat.

Appendix G-II-D-4-e. Laboratory furniture is of simple and sturdy construction, and spaces between benches, cabinets, and equipment are accessible for cleaning.

Appendix G-II-D-4-f. A foot, elbow, or automatically operated hand-washing sink is provided near the door of each laboratory room in the facility.

Appendix G-II-D-4-g. If there is a central vacuum system, it does not serve areas outside the facility. In-line HEPA filters are placed as near as practicable to each use point or service cock. Filters are installed to permit in-place decontamination and replacement. Other liquid and gas services to the facility are protected by devices that prevent backflow.

Appendix G-II-D-4-h. If water fountains are provided, they are foot operated and are located in the facility corridors outside the laboratory. The water service to the fountain is not connected to the backflow-protected distribution system supplying water to the laboratory areas.

Appendix G-II-D-4-i. Access doors to the laboratory are self-closing and lockable.

Appendix G-II-D-4-j. Any windows are breakage resistant.

Appendix G-II-D-4-k. A double-doored autoclave is provided for decontaminating materials passing out of the facility. The autoclave door which opens to the area external to the facility is sealed to the outer wall and automatically controlled so that the outside door can only be opened after the autoclave "sterilization" cycle has been completed.

Appendix G-II-D-4-l. A pass-through dunk tank, fumigation chamber, or an equivalent decontamination method is provided so that materials and equipment that cannot be decontaminated in the autoclave can be safely removed from the facility.

Appendix G-II-D-4-m. Liquid effluents from laboratory sinks, biological safety cabinets, floors, and autoclave chambers are decontaminated by heat treatment before being released from the maximum containment facility. Liquid wastes from shower rooms and toilets may be decontaminated with chemical disinfectants or by heat in the liquid waste decontamination system. The procedure used for heat decontamination of liquid wastes is evaluated mechanically and biologically by using a recording thermometer and an indicator microorganism with a defined heat susceptibility pattern. If liquid wastes from the shower room are decontaminated with chemical disinfectants, the chemical used is of demonstrated efficacy against the target or indicator microorganisms.

Appendix G-II-D-4-n. An individual supply and exhaust air ventilation system is provided. The system maintains pressure differentials and directional airflow as required to assure flows inward from areas outside of the facility toward areas of highest potential risk within the facility. Manometers are used to sense pressure differentials between adjacent areas maintained at different pressure levels. If a system malfunctions, the manometers sound an alarm. The supply and exhaust airflow is interlocked to assure inward (or zero) airflow at all times.

Appendix G-II-D-4-o. The exhaust air from the facility is filtered through HEPA filters and discharged to the outside so that it is dispersed away from occupied buildings and air intakes. Within the facility, the filters are located as near the laboratories as practicable in order to reduce the length of potentially contaminated air ducts. The filter chambers are designed to allow *in situ* decontamination before filters are removed and to facilitate certification testing after they are replaced. Coarse filters and HEPA filters are provided to treat air supplied to the facility in order to increase the lifetime of the exhaust HEPA filters and to protect the supply air system should air pressures become unbalanced in the laboratory.

Appendix G-II-D-4-p. The treated exhaust air from Class I and II biological safety cabinets can be discharged into the laboratory room environment or the outside through the facility air exhaust system. If exhaust air from Class I or II biological safety cabinets is discharged into the laboratory the cabinets are tested and certified at 6-month intervals. *The treated exhaust air from Class III biological safety cabinets is discharged, without recirculation through two sets of HEPA filters in series, via the facility exhaust air system.* If the treated exhaust air from any of these cabinets is discharged to the outside through the facility exhaust air system, it is connected to this system in a manner (e.g., thimble unit connection (12)) that avoids any interference with the air balance of the cabinets or the facility exhaust air system.

Appendix G-II-D-4-q. A specially designed suit area may be provided in the facility. Personnel who enter this area wear a one-piece positive pressure suit that is ventilated by a life-support system. The life-support system includes alarms and emergency backup breathing air tanks. Entry to this area is through an airlock fitted with airtight doors. A chemical shower is provided to decontaminate the surface of the suit before the worker leaves the area. The exhaust air from the suit area is filtered by two sets of HEPA filters installed in series. A duplicate filtration unit, exhaust fan, and an automatically starting emergency power source are provided. The air pressure within the suit area is lower than that of any adjacent area. Emergency lighting and communication systems are provided. All penetrations into the internal shell of the suit area are sealed. A double-doored autoclave is provided for decontaminating waste materials to be removed from the suit area.

Appendix G-III.   *Footnotes and References of Appendix G*

(1)   Laboratory Safety at the Center for Disease Control (Sept. 1974), U.S. Department of Health, Education and Welfare Publication No. CDC 7508118.

(2)   *Biosafety in Microbiological and Biomedical Laboratories.* 1st Edition (March 1984), U.S. Department of Health and Human Services, Public Health Service, Centers for Disease Control, Atlanta, Georgia 30333, and National Institutes of Health, Bethesda, Maryland 20205.

(3)   *National Cancer Institute Safety Standards for Research Involving Oncogenic Viruses* (Oct. 1974), U.S. Department of Health, Education and Welfare Publication No. (NIH) 75-790.

(4)   *National Institutes of Health Biohazards Safety Guide* (1974), U.S. Department of Health, Education and Welfare, Public Health Service, National Institutes of Health, U.S. Government Printing Office, Stock No. 1740-00383.

(5)   *Biohazards in Biological Research* (1973). A. Hellman, M. N. Oxman, and R. Pollack (ed.). Cold Spring Harbor Laboratory.

(6)   *Handbook of Laboratory Safety* (1971). 2nd Edition, N. V. Steere (ed.). The Chemical Rubber Co., Cleveland.

(7)   Bodily, J. L. (1970). *General Administration of the Laboratory,* H. L. Bodily, E. L. Updyke, and J. O. Mason (eds.). Diagnostic Procedures for Bacterial, Mycotic and Parasitic Infections. American Public Health Association, New York, pp. 11-28.

(8) Darlow, H. M. (1969). *Safety in the Microbiological Laboratory*. In J. R. Norris and D. W. Robbins (ed.). Methods in Microbiology, Academic Press, Inc., New York, pp. 169-204.

(9) *The Prevention of Laboratory Acquired Infection* (1974). C. H. Collins, E. G. Hartley, and R. Pilsworth, Public Health Laboratory Service, Monograph Series No. 6.

(10) Chatigny, M. A. (1961). *Protection Against Infection in the Microbiological Laboratory: Devices and Procedures*. In W. W. Umbreit (ed.). Advances in Applied Microbiology, Academic Press, New York, N.Y. 3:131-192.

(11) Horsfall, F. L., Jr., and J. H. Baner (1940). *Individual Isolation of Infected Animals in a Single Room*. J. Bact. 40, 569-580.

(12) Biological safety cabinets referred to in this section are classified as *Class I, Class II,* or *Class III* cabinets. A *Class I* is a ventilated cabinet for personnel protection having an inward flow of air away from the operator. The exhaust air from this cabinet is filtered through a high-efficiency particulate air (HEPA) filter. This cabinet is used in three operational modes: (1) With a full-width open front, (2) with an installed front closure panel (having four 8-inch diameter openings) without gloves, and (3) with an installed front closure panel equipped with arm-length rubber gloves. The face velocity of the inward flow of air through the full-width open front is 75 feet per minute or greater.

A *Class II* cabinet is a ventilated cabinet for personnel and product protection having an open front with inward air flow for personnel protection, and HEPA filtered mass recirculated air flow for product protection. The cabinet exhaust air is filtered through a HEPA filter. The face velocity of the inward flow of air through the full-width open front is 75 feet per minute or greater. Design and performance specifications for *Class II* cabinets have been adopted by the National Sanitation Foundation, Ann Arbor, Michigan. A *Class III* cabinet is a closed front ventilated cabinet of gas-tight construction which provides the highest level of personnel protection of all biohazard safety cabinets. The interior of the cabinet is protected from contaminants exterior to the cabinet. The cabinet is fitted with arm-length rubber gloves and is operated under a negative pressure of at least 0.5 inches water gauge. All supply air is filtered through HEPA filters. Exhaust air is filtered through two HEPA filters or one HEPA filter and incinerator before being discharged to the outside environment. National Sanitation Foundation Standard 49. 1976. Class II (Laminar Flow) Biohazard Cabinetry, Ann Arbor, Michigan.

(13) Biosafety Level 1 is suitable for work involving agents of no known or minimal potential hazard to laboratory personnel and the environment. The laboratory is not separated from the general traffic patterns in the building. Work is generally conducted on open bench tops. Special containment equipment is not required or generally used. Laboratory personnel have specific training in the procedures conducted in the laboratory and are supervised by a scientist with general training in microbiology or a related science (see Appendix G-III-2).

(14) Biosafety Level 2 is similar to Level 1 and is suitable for work involving agents of moderate potential hazard to personnel and the environment. It differs in that: (1) Laboratory personnel have specific training in handling pathogenic agents and are directed by competent scientists; (2) access to the laboratory is limited when work is being conducted; and (3) certain procedures in which infectious aerosols are created are conducted in biological safety cabinets or other physical containment equipment (see Appendix G-III-2).

(15) Office of Research Safety, National Cancer Institute, and the Special Committee of Safety and Health Experts, 1978, "Laboratory Safety Monograph: A Supplement to the NIH Guidelines for Recombinant DNA Research". Bethesda, Maryland, National Institutes of Health.

(16) Biosafety Level 3 is applicable to clinical, diagnostic, teaching, research, or production facilities in which work is done with indigenous or exotic agents which may cause serious or potentially lethal disease as a result of exposure by the inhalation route. Laboratory personnel have specific training in handling pathogenic and potentially lethal agents and are supervised by competent scientists who are experienced in working with these agents. All procedures involving the manipulation of infectious material are conducted within biological safety cabinets or other physical containment devices or by personnel wearing appropriate personal protective clothing and devices. The laboratory has special engineering and design features. It is recognized, however, that many existing facilities may not have all the facility safeguards recommended for Biosafety Level 3 (e.g., access zone, sealed penetrations, and directional airflow, etc.). In these circumstances, acceptable safety may be achieved for routine or repetitive operations (e.g., diagnostic procedures involving the propagation of an agent for identification, typing, and susceptibility testing) in laboratories where facility features satisfy Biosafety Level 2 recommendations provided the

recommended "Standard Microbiological Practices", "Special Practices", and "Containment Equipment" for Biosafety Level 3 are rigorously followed. The decision to implement this modification of Biosafety Level 3 recommendations should be made only by the laboratory director (see Appendix G-III-2).

Appendix H. Shipment

Recombinant DNA molecules contained in an organism or virus shall be shipped only as an etiologic agent under requirements of the U.S. Public Health Service, and the

PACKAGING AND LABELING OF ETIOLOGIC AGENTS

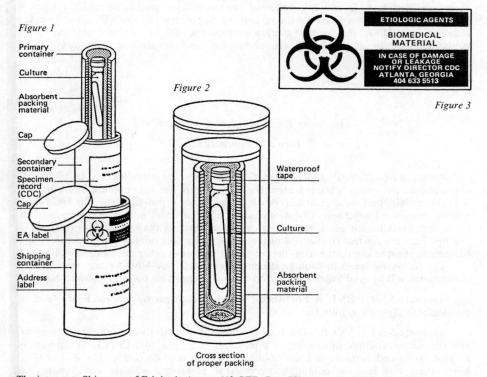

Figure 1

Primary container

Culture

Absorbent packing material

Cap

Secondary container

Specimen record (CDC)

Cap

EA label

Shipping container

Address label

Figure 2

Waterproof tape

Culture

Absorbent packing material

ETIOLOGIC AGENTS
BIOMEDICAL MATERIAL
IN CASE OF DAMAGE OR LEAKAGE NOTIFY DIRECTOR CDC ATLANTA, GEORGIA 404 633 5513

Figure 3

Cross section of proper packing

The interstate Shipment of Etiologic Agents (42 CFR, Part 72) was revised July 21, 1980 to provide for packaging and labeling requirements for etiologic agents and certain other materials shipped in interstate traffic.

Figures 1 and 2 diagram the packaging and labeling of etiologic agents in volumes of less than 50 ml. in accordance with the provisions of subparagraph 72.3 *(a)* of the cited regulation. Figure 3 illustrates the size of the label, described in subparagraph 72.3 *(d)* (1-5) of the regulations, which shall be affixed to all shipments of etiologic agents.

For further information on any provision of this regulation contact:

Centers for Disease Control
Attn: Biohazards Control Office
1600 Clifton Road
Atlanta, Georgia 30333

Telephone: 404-329-3883
FTS-236-3883

U.S. Department of Transportation (Section 72.3, Part 72, Title 42, and Sections 173.386-.388, Part 173, Title 49, U.S. Code of Federal Regulations (CFR)) as specified below:

Appendix H-I. Recombinant DNA molecules contained in an organism or virus requiring BL1, BL2, or BL3 physical containment, when offered for transportation or transported, are subject to all requirements of Section 72.3(a)-(e), Part 72, Title 42, CFR, and Sections 173.386-.388, Part 173, Title 49 CFR.

Appendix H-II. Recombinant DNA molecules contained in an organism or virus requiring BL4 physical containment, when offered for transportation or transported, are subject to the requirements listed above under Appendix H-I and are also subject to Section 72.3(f), Part 72, Title 42 CFR.

Appendix H-III. Information on packaging and labelling of etiologic agents is shown in Figures 1, 2, and 3. Additional information on packaging and shipment is given in the "Laboratory Safety Monograph—A Supplement to the NIH Guidelines for Recombinant DNA Research", available from the Office of Recombinant DNA Activities and in *Biosafety in Microbiological and Biomedical Laboratories* (see Appendix G-III-2).

Appendix I.   Biological Containment

(See also Appendix E.)

Appendix I-I.   Levels of Biological Containment. In consideration of biological containment, the vector (plasmid, organelle, or virus) for the recombinant DNA and the host (bacterial, plant, or animal cell) in which the vector is propagated in the laboratory will be considered together. Any combination of vector and host which is to provide biological containment must be chosen or constructed so that the following types of "escape" are minimized: (i) survival of the vector in its host outside the laboratory, and (ii) transmission of the vector from the propagation host to other nonlaboratory hosts.

The following levels of biological containment (HV, or Host-Vector, systems) for prokaryotes will be established; specific criteria will depend on the organisms to be used.

Appendix I-I-A. HV1. A host-vector system which provides a moderate level of containment. *Specific system* are:

Appendix I-I-A-1.   EK1. The host is always *E. coli* K-12 or a derivative thereof, and the vectors include nonconjugative plasmids (e.g., pSC101, ColE1, or derivatives thereof (1-7)) and variants of bacteriophage, such as lambda (8-15). The *E. coli* K-12 hosts shall not contain conjugation-proficient plasmids, whether autonomous or integrated, or generalized tranducing phages.

Appendix I-I-A-2.   Other HV1. Hosts and vectors shall be, at a minimum, comparable in containment to *E. coli* K-12 with a non conjugative plasmid or bacteriophage vector. The data to be considered and a mechanism for approval of such HV1 systems are described below (Appendix I-II).

Appendix I-I-B.   HV2. These are host-vector systems shown to provide a high level of biological containment as demonstrated by data from suitable tests performed in the laboratory. Escape of the recombinant DNA either via survival of the organisms or via transmission of recombinant INA to other organisms should be less than 1/10 G58 under specified conditions. Specific systems are:

Appendix I-I-B-1. For EK2 host-vector systems in which the vector is a plasmid, no more than one in . . . host cells should be able to perpetuate a cloned DNA fragment under the specified nonpermissive laboratory conditions designed to represent the natural environment, either by survival of the original host or as a consequence of transmission of the cloned DNA fragment.

Appendix I-I-B-2. For EK2 host-vector systems in which the vector is a phage, no more than one in . . . phage particles should be able to perpetuate a cloned DMA fragment under the specified nonpermissive laboratory conditions designed to represent the natural environment either: (i) as a prophage (in the inserted or plasmid form) in the laboratory host used for phage propagation or (ii) by surviving in natural environments and transferring a cloned DNA fragment to other hosts (or their resident prophages).

Appendix I-II. Certification of Host-Vector Systems.

Appendix I-II-A. Responsibility. HV1 systems other than *E. coli* K-12 and HV2 host-vector systems may not be designated as such until they have been certified by the Director, NIH. Application for certification of a host-vector system is made by written application to the Office of Recombinant DNA Activities, National Institutes of Health, Building 31, Room 3B10, Bethesda, Maryland 20205.

Host-vector systems that are proposed for certification will be reviewed by the RAC (see Section IV-C-1-b-(1)-(e)). This will first involve review of the data on construction, properties, and testing of the proposed host-vector system by a working group composed of one or more members of the RAC and other persons chosen because of their expertise in evaluating such data. The committee will then evaluate the report of the working group and any other available information at a regular meeting. The Director, NIH, is responsible for certification after receiving the advice of the RAC. Minor modifications of existing certified Host-vector systems where the modifications are of minimal or no consequence to the properties relevant to containment may be certified by the Director, NIH, without review by the RAC (see Section IV-C-1-b-(3)-(c)).

When new host-vector systems are certified, notice of the certification will be sent by ORDA to the applicant and to all IBCs and will be published in the *Recombinant DNA Technical Bulletin.* Copies of a list of all currently certified host-vector systems may be obtained from ORDA at any time.

The Director, NIH, may at any time rescind the certification of any host-vector system (see Section IV-C-1-b-(3)-(d)). If certification of a host-vector system is rescinded, NIH will instruct investigators to transfer cloned DNA into a different system or use the clones at a higher physical containment level unless NIH determines that the already constructed clones incorporate adequate biological containment.

Certification of a given system does not extend to modifications of either the host or vector component of that system. Such modified systems must be independently certified by the Director, NIH. If modifications are minor, it may only be necessary for the investigator to submit data showing that the modifications have either improved or not impaired the major phenotypic traits on which the containment of the system depends. Substantial modifications of a certified system require the submission of complete testing data.

Appendix I-II-B. Data to be Submitted for Certification.

Appendix I-II-B-1. HV1 Systems Other than E. coli *K-12.* The following types of data shall be submitted, modified as appropriate for the particular system under consideration: (i) a description of the organism and vector; the strain's natural habitat and growth requirements; its physiological properties, particularly those related to its reproduction and survival and the mechanisms by which it exchanges genetic information; the range of organisms with which this organism normally exchanges genetic information and what sort of information is exchanged; and any relevant information on its pathogenicity or toxicity; (ii) a description of the history of the particular strains and vectors to be used, including data on any mutations which render this organism less able to survive or transmit genetic information; and (iii) a general description of the range of experiments contemplated with emphasis on the need for developing such an HV1 system.

Appendix I-II-B-2. HV2 Systems. Investigators planning to request HV2 certification for host-vector systems can obtain instructions from ORDA concerning data to be submitted (14-15). In general, the following types of data are required: (i) description of construction steps with indication of source, properties, and manner of introduction of genetic traits; (ii) quantitative data on the stability of genetic traits that contribute to the containment of the system; (iii) data on the survival of the host-vector system under nonpermissive laboratory conditions designed to represent the relevant natural environment; (iv) data on transmissibility of the vector and/or a cloned DNA fragment under both permissive and nonpermissive conditions; (v) data on all other properties of the system which affect containment and utility, including information on yields of phage or plasmid molecules, ease of DNA isolation, and ease of transfect ion or transformation; and (vi) in some cases the investigator may be asked to submit data on survival and vector transmissibility from experiments in which the host-vector is fed to laboratory animals and human subjects. Such *in vivo* data may be required to confirm the validity of predicting *in vivo* survival on the basis of *in vitro* experiments.

Data must be submitted in writing to ORDA. Ten to twelve weeks are normally required for review and circulation of the data prior to the meeting at which such data can be considered by the RAC. Investigators are encouraged to publish their data on the construction, properties, and testing of proposed HV2 systems prior to consideration of the system by the RAC and its subcommittee. More specific instructions concerning the type of data to be submitted to NIH for proposed EK2 systems involving either plasmids or bacteriophage in *E. coli* K-12 are available from ORDA.

Appendix I-III. Footnotes and References of Appendix I

(1) Hershfield, V., H. W. Boyer, C. Yanofsky, M. A. Lovett, and D. R. Helinski (1974). *Plasmid ColEl as a Molecular Vehicle for Cloning and Amplification of DNA.* Proc. Nat. Acad. Sci. USA 71, 3455-3459.

(2) Wensink, P.C., D.J. Finnegan, J.E. Donelson, and D.S. Hogness (1974). *A System for Mapping DNA Sequences in the Chromosomes of Drosophila Melanogaster.* Cell 3, 315-335.

(3) Tanaka, T., and B. Weisblum (1975). *Construction of a Colicin El-R Factor Composite Plasmid in Vitro: Means for Amplification of Deoxyribonucleic Acid.* J. Bacteriol. 121. 354-382.

(4) Armstrong, K. A., V. Hershfield, and D. R. Helinski (1977). *Gene Cloning and Containment Properties of Plasmid Col El and its Derivatives,* Science 196, 172-174.

(5) Bolivar, F., R. L. Rodriguez, M. C. Betlach, and H. W. Boyer (1977). *Construction and Characterization of New Cloning Vehicles: I. Ampicillin-Resistant Derivative of pMB9.* Gene 2, 75-93.

(6) Cohen, S. N., A. C. W. Chang, H. Boyer, and R. Helling (1973). *Construction of Biologically Functional Bacterial Plasmids in Vitro.* Proc. Natl. Acad. Sci. USA 70 3240-3244.

(7) Bolivar, F., R. L. Rodriguez, R. J. Greene, M. C. Batlach, H. L. Reyneker, H. W. Boyer, J. H. Crosa, and S. Falkow (1977). *Construction and Characterization of New Cloning Vehicles: II. A Multi-Purpose Cloning System.* Gene 2, 95-113.

(8) Thomas, M., J. R. Cameron, and R. W. Davis (1974). *Viable Molecular Hybrids of Bacteriophage Lambda and Eukaryotic DNA.* Proc. Nat. Acad. Sci. USA 71, 4579-4583.

(9) Murray, E., and K. Murray (1974). *Manipulation of Restriction Targets in Phage Lambda to Farm Receptor Chromosomes for DNA Fragments.* Nature 251, 476-481.

(10) Rambach A., and P. Tiollais (1974). *Bacteriophage Having EcoRI Endonuclease Sites Only in the Non-Essential Region of the Genome.* Proc. Nat. Acad. Sci., USA 71, 3927-3930.

(11) Blattner, F. R., B. G. Williams, A. E. Bleche, K. Denniston-Thompson, H. E. Faber, I. A. Furlong, D. J. Gunwald, D. O. Kiefer, D. D. Moore, J. W. Shumm, E. L. Sheldon, and O. Smithies (1977). *Charon Phages; Safer Derivatives of Bacteriophage Lambda for DNA cloning.* Science 196, 163-169.

(12) Donoghue, D. J., and P. A. Sharp (1977). *An Improved Lambda Vector: Construction of Model Recombinants Coding for Kanamycin Resistance.* Gene 1, 209-227.

(13) Leder, P., D. Tiemeier and L. Enquist (1977). *EK2 Derivatives of Bacteriophage Lambda Useful in the Cloning of DNA from Higher Organisms: The ???????? WES System.* Science 196, 175-177.

(14) Skalka, A. (1978). *Current Status of Coliphage, EK2 Vectors.*

(15) Szybalski, W., A. Skalka, S. Gottesman, A. Campbell, and D. Botstein (1978). *Standardized Laboratory Tests for EK2 Certification.* Gene 3, 36-38.

Appendix J. Federal Interagency Advisory Committee on Recombinant DNA Research

Appendix J-I. Federal Interagency Advisory Committee. The Federal Interagency Advisory Committee on Recombinant DNA Research advises the Secretary of the Department of Health and Human Services, the Assistant Secretary for Health, and the Director, National Institutes of Health, on the coordination of those aspects of all Federal programs and activities relating to recombinant DNA research. The committee provides for communication and exchange of information necessary to maintain adequate coordination of such programs and activities. The committee is responsible for facilitating compliance with a uniform set of guidelines in the conduct of this research in the public and private sectors and, where warranted, to suggest administrative or legislative proposals.

The Director of the NIH, or his designee, serves as chairman, and the committee includes representation from all Departments and Agencies whose programs involve health functions or responsibilities as determined by the Secretary.

Departments and agencies which have representation on this committee as of December 1980 are:

Department of Agriculture
Department of Commerce
Department of Defense
Department of Energy
Environmental Protection Agency
Executive Office of the President
Department of Health and Human Services
Office of the Assistant Secretary for Health Centers for Disease Control
Food and Drug Administration
National Institutes of Health
Department of the Interior
Department of Justice
Department of Labor
National Aeronautics and Space Administration
National Science Foundation
Nuclear Regulatory Commission
Department of State
Department of Transportation
Arms Control and Disarmament Agency Veterans Administration

At the second meeting of the committee on November 23, 1976, all of the Federal agencies endorsed the Guidelines, and Departments which support or conduct recombinant DNA research agreed to abide by the Guidelines (1).

(1)   Minutes of the first eight meetings of the Federal Interagency Advisory Committee on Recombinant DNA Research are reproduced in *Recombinant DNA Research, Volume 2. Documents Relating to "NIH Guidelines for Research Involving Recombinant DNA Molecules," June 1976-November 1977.*

Apendix K.   Physical Containment for Large-Scale Uses of Organisms Containing Recombinant DNA Molecules

This part of the Guidelines specifies physical containment guidelines for large-scale (greater than 10 liters of culture) research or production involving viable organisms containing recombinant DNA molecules. It shall apply to large-scale research or production activities as specified in Section III-B-5 of the Guidelines.

All provisions of the Guidelines shall apply to large-scale research or production activities with the following modifications:

●   Appendix K shall replace Appendix G when quantities in excess of 10 liters of culture are involved in research or production.

●   The institution shall appoint a Biological Safety Officer (BSO) if it engages in large-scale research or production activities involving viable organisms containing recombinant DNA molecules. The duties of the BSO shall include those specified in Section IV-B-4 of the Guidelines.

●   The institution shall establish and maintain a health surveillance program for personnel engaged in large-scale research or production activities involving viable organisms containing recombinant DNA molecules which require BL3 containment at the laboratory scale. The program shall include: preassignment and periodic physical and medical examinations; collection, maintenance and analysis of serum specimens for monitoring serologic changes that may result from the employee's work experience; and provisions for the investigation of any serious, unusual or extended illnesses of employees to determine possible occupational origin.

Appendix K-I.   Selection of Physical Containment Levels.   The selection of the physical containment level required for recombinant DNA research or production involving more than 10 liters of culture is based on the containment guidelines established in Part III of the Guidelines. For purposes of large-scale research or production, three physical containment levels are established. These are referred to as BL1-LS, BL2-LS, and BL3-LS. The BL-LS level of physical containment is required for large-scale research or production of viable organisms containing recombinant DNA molecules which require BL1 containment at the laboratory scale. (The BL1-LS level of physical containment is recommended for large-scale research or production of viable organisms for which BL1 is recommended at the laboratory scale such as those described in Appendix C.) The BL2-LS level of physical containment is required for large-scale research or production of viable organisms containing recombinant DNA molecules which require BL2 containment at the laboratory scale. The BL3-LS level of physical containment is required for large-scale research or production of viable organisms containing recombinant DNA molecules which require BL3 containment at the laboratory scale. No provisions are made for large-scale research or production of viable organisms containing recombinant DNA molecules which require BL4 containment at the laboratory scale. If necessary, these requirements will be established by NIH on an individual basis.

Appendix K-II. BL1-LS Level

Appendix K-II-A. Cultures of viable organisms containing recombinant DNA molecules shall be handled in a closed system (e.g., closed vessel used for the propagation and growth of cultures) or other primary containment equipment (e.g., biological safety cabinet containing a centrifuge used to process culture fluids) which is designed to reduce the potential for escape of viable organisms. Volumes less than 10 liters may be handled outside of a closed system or other primary containment equipment provided all physical containment requirements specified in Appendix C-II-A of the Guidelines are met.

Appendix K-II-B. Culture fluids (except as allowed in Appendix K-II-C) shall not be removed from a closed system or other primary containment equipment unless the viable organisms containing recombinant DNA molecules have been inactivated by a validated inactivation procedure. A validated inactivation procedure is one which has been demonstrated to be effective using the organism that will serve as the host for propagating the recombinant DNA molecules.

Appendix K-II-C. Sample collection from a closed system, the addition of materials to a closed system, and the transfer of culture fluids from one closed system to another shall be done in a manner which minimizes the release of aerosols or contamination of exposed surfaces.

Appendix K-II-D. Exhaust gases removed from a closed system or other primary containment equipment shall be treated by filters which have efficiencies equivalent to HEPA filters or by other equivalent procedures (e.g., incineration) to minimize the release of viable organisms containing recombinant DNA molecules to the environment.

Appendix K-II-E. A closed system or other primary containment equipment that has contained viable organisms containing recombinant DNA molecules shall not be opened for maintenance or other purposes unless it has been sterilized by a validated sterilization procedure. A validated sterilization procedure is one which has been demonstrated to be effective using the organism that will serve as the host for propagating the recombinant DNA molecules.

Appendix K-II-F. Emergency plans required by Section IV-B-3-f shall include methods and procedures for handling large losses of culture on an emergency basis.

Appendix K-III. BL2-LS Level

Appendix K-III-A. Cultures of viable organisms containing recombinant DNA molecules shall be handled in a closed system (e.g., closed vessel used for the propagation and growth of cultures) or other primary containment equipment (e.g., Class III biological safety cabinet containing a centrifuge used to process culture fluids) which is designed to prevent the escape of viable organisms. Volumes less than 10 liters may be handled outside of a closed system or other primary containment equipment provided all physical containment requirements specified in Appendix C-II-B of the Guidelines are met.

Appendix K-III-B. Culture fluids (except as allowed in Appendix K-III-C) shall not be removed from a closed system or other primary containment equipment unless the viable organisms containing recombinant DNA molecules have been inactivated by a validated inactivation procedure. A validated inactivation procedure is one which has been demonstrated to be effective using the organism that will serve as the host for propagating the recombinant DNA molecules.

Appendix K-III-C. Sample collection from a closed system, the addition of materials to a closed system, and the transfer of cultures fluids from one closed system to another shall be done in a manner which prevents the release of aerosols or contamination of exposed surfaces.

Appendix K-III-D. Exhaust gases removed from a closed system or other primary containment equipment shall be treated by filters which have efficiencies equivalent to HEPA filters or by other equivalent procedures (e.g., incineration) to prevent the release of viable organisms containing recombinant DNA molecules to the environment.

Appendix K-III-E. A closed system or other primary containment equipment that has contained viable organisms containing recombinant DNA molecules shall not be opened for maintenance or other purposes unless it has been sterilized by a validated sterilization procedure. A validated sterilization procedure is one which has been demonstrated to be effective using the organism that will serve as the host for propagating the recombinant DNA molecules.

Appendix K-III-F. Rotating seals and other mechanical devices directly associated with a closed system used for the propagation and growth of viable organisms containing recombinant DNA molecules shall be designed to prevent leakage or shall be fully enclosed in ventilated housings that are exhausted through filters which have efficiencies equivalent to HEPA filters or through other equivalent treatment devices.

Appendix K-III-G. A closed system used for the propagation and growth of viable organisms of containing recombinant DNA molecules and other primary containment equipment used to contain operations involving viable organisms containing recombinant DNA molecules shall include monitoring or sensing devices that monitor the integrity of containment during operations.

Appendix K-III-H. A closed system used for the propagation and growth of viable organisms containing recombinant DNA molecules shall be tested for integrity of the containment features using the organism that will serve as the host for propagating recombinant DNA molecules. Testing shall be accomplished prior to the introduction of viable organisms containing recombinant DNA molecules and following modification or replacement of essential containment features. Procedures and methods used in the testing shall be appropriate for the equipment design and for recovery and demonstration of the test organism. Records of tests and results shall be maintained on file.

Appendix K-III-I. A closed system used for the propagation and growth of viable organisms containing recombinant DNA molecules shall be permanently identified. This identification shall be used in all records reflecting testing, operation, and maintenance and in all documentation relating to use of this equipment for research or production activities involving viable organisms containing recombinant DNA molecules.

Appendix K-III-J. The universal biohazard sign shall be posted on each closed system and primary containment equipment when used to contain viable organisms containing recombinant DNA molecules.

Appendix K-III-K. Emergency plans required by Section IV-B-3-f shall include methods and procedures for handling large losses of culture on an emergency basis.

Appendix K-IV. BL3-LS Level.

Appendix K-IV-A. Cultures of viable organisms containing recombinant DNA molecules shall be handled in a closed system (e.g., closed vessels used for the propagation and growth of cultures) or other primary containment equipment (e.g., Class III biological safety cabinet containing a centrifuge used to process culture fluids) which is designed to prevent the escape of viable organisms. Volumes less than 10 liters may be handled outside of a closed system provided all physical containment requirements specified in Appendix G-II-C of the Guidelines are met.

Appendix K-IV-B. Culture fluids (except as allowed in Appendix K-IV-C) shall not be removed from a closed system or other primary containment equipment unless the viable organisms containing recombinant DNA molecules have been inactivated by a validated inactivation procedure. A validated inactivation procedure is one which has been demonstrated to be effective using the organisms that will serve as the host for propagating the recombinant DNA molecules.

Appendix K-IV-C. Sample collection from a closed system, the addition of materials to a closed system and the transfer of culture fluids from one closed system to another shall be done in a manner which prevents the release of aerosols or contamination of exposed surfaces.

Appendix K-IV-D. Exhaust gases removed from a closed system or other primary containment equipment shall be treated by filters which have efficiencies equivalent to HEPA filters or by other equivalent procedures (e.g., incineration to prevent the release of viable organisms containing recombinant DNA molecules to the environment).

Appendix K-IV-E. A closed system or other primary containment equipment that has contained viable organisms containing recombinant DNA molecules shall not be opened for maintenance or other purposes unless it has been sterilized by a validated sterilization procedure. A validated sterilization procedure is one which has been demonstrated to be effective using the organisms that will serve as the host for propagating the recombinant DNA molecules.

Appendix K-IV-F. A closed system used for the propagation and growth of viable organisms containing recombinant DNA molecules shall be operated so that the space above the culture level will be maintained at a pressure as low as possible, consistent with equipment design, in order to maintain the integrity of containment features.

Appendix K-IV-G. Rotating seals and other mechanical devices directly associated with a closed system used to contain viable organisms containing recombinant DNA molecules shall be designed to prevent leakage or shall be fully enclosed in ventilated housings that are exhausted through filters which have efficiencies equivalent to HEPA filters or through other equivalent treatment devices.

Appendix K-IV-H. A closed system used for the propagation and growth of viable organisms containing recombinant DNA molecules and other primary containment equipment used to contain operations involving viable organisms containing recombinant DNA molecules shall include monitoring or sensing devices that monitor the integrity of containment during operations.

Appendix K-IV-I. A closed system used for the propagation and growth of viable organisms containing recombinant DNA molecules shall be tested for integrity of the containment features using the organisms that will serve as the host for propagating the recombinant DNA molecules. Testing shall be accomplished prior to the introduction of viable organisms containing recombinant DNA molecules and following modification or replacement of essential containment features. Procedures and methods used in the testing shall be appropriate for the equipment design and for recovery and demonstration of the test organism. Records of tests and results shall be maintained on file.

Appendix K-IV-J. A closed system used for the propagation and growth of viable organisms containing recombinant DNA molecules shall be permanently identified. This identification shall be used in all records reflecting testing, operation, and maintenance and in all documentation relating to the use of this equipment for research production activities involving viable organisms containing recombinant DNA molecules.

Appendix K-IV-K. The universal biohazard sign shall be posted on each closed system and primary containment equipment when used to contain viable organisms containing recombinant DNA molecules.

Appendix K-IV-L. Emergency plans required by Section IV-B-3-f shall include metods and procedures for handling large losses of culture on an emergency basis.

Appendix K-IV-M. Closed systems and other primary containment equipment used in handling cultures of viable organisms containing recombinant DNA molecules shall be located within a controlled area which meets the following requirements:

Appendix K-IV-M-1. The controlled area shall have a separate entry area. The entry area shall be a double-doored space such as an air lock, anteroom, or change room that separates the controlled area from the balance of the facility.

Appendix K-IV-M-2. The surfaces of walls, ceilings, and floors in the controlled area shall be such as to permit ready cleaning and decontamination.

Appendix K-IV-M-3. Penetrations into controlled area shall be sealed to permit liquid or vapor phase space decontamination.

Appendix K-IV-M-4. All utilities and service or process piping and wiring entering the controlled area shall be protected against contamination.

Appendix K-IV-M-5. Hand-washing facilities equipped with foot, elbow, or automatically operated valves shall be located at each major work area and near each primary exit.

Appendix K-IV-M-6. A shower facility shall be provided. This facility shall be located in close proximity to the controlled area.

Appendix K-IV-M-7. The controlled area shall be designed to preclude release of culture fluids outside the controlled area in the event of an accidental spill or release from the closed systems or other primary containment equipment.

Appendix K-IV-M-8. The controlled area shall have a ventilation system that is capable of controlling air movement. The movement of air shall be from areas of lower contamination potential to areas of higher contamination potential. If the ventilation system provides positive pressure supply air, the system shall operate in a manner that prevents the reversal of the direction of air movement or shall be equipped with an alarm that would be actuated in the event that reversal in the direction of air movement were to occur. The exhaust air from the controlled area shall not be recirculated to other areas of the facility. The exhaust air from the controlled area may be discharged to the outdoors without filtration or other means for effectively reducing an accidental aerosol burden provided that it can be dispersed clear of occupied buildings and air intakes.

Appendix K-IV-N-. The following personnel and operational practices shall be required:

Appendix K-IV-N-1. Personnel entry into the controlled area shall be through the entry area specified in Appendix K-IV-M-1.

Appendix K-IV-N-2. Persons entering the controlled area shall exchange or cover their personal clothing with work garments such as jumpsuits, laboratory coats, pants and shirts, ear cover and shoes or shoe covers. On exit from the controlled area the work clothing may be stored in a locker separate from that used for personal clothing or discarded for laundering. Clothing shall be decontaminated before laundering.

Appendix K-IV-N-3. Entry into the controlled area during periods when work is in progress shall be restricted to those persons required to meet program or support needs. Prior to entry all persons shall be informed of the operating practices, emergency procedures, and the nature of the work conducted.

Appendix K-IV-N-4. Persons under 18 years shall not be permitted to enter the controlled area.

Appendix K-IV-N-5. The universal biohazard sign shall be posted on entry doors to the controlled area and all internal doors when any work involving the organism is in progress. This includes periods when decontamination procedures are in progress. The sign posted on the entry doors to the controlled area shall include a statement of agents in use and personnel authorized to enter the controlled area.

Appendix K-IV-N-6. The controlled area shall be kept neat and clean.

Appendix K-IV-N-7. Eating, drinking, smoking, and storage of food are prohibited in the controlled area.

Appendix K-IV-N-8. Animals and plants shall be excluded from the controlled area.

Appendix K-IV-N-9. An effective insect and rodent control program shall be maintained.

Appendix K-IV-N-10. Access doors to the controlled area shall be kept closed, except as necessary for access, while work is in progress. Serve doors leading directly outdoors shall be sealed and locked while work is in progress.

Appendix K-IV-N-11. Persons shall wash their hands when leaving the controlled area.

Appendix K-IV-N-12. Persons working in the controlled area shall be trained in emergency procedures.

Appendix K-IV-N-13. Equipment and materials required for the management of accidents involving viable organisms containing recombinant DNA molecules shall be available in the controlled area.

Appendix K-IV-N-14. The controlled area shall be decontaminated in accordance with established procedures following spills or other accidental release of viable organisms containing recombinant DNA molecules.

Appendix L.   Release Into the Environment of Certain Plants

Appendix L-1.   General Information. Appendix L specifies conditions under which certain plants as specified below, may be approved for release into the environment. Experiments in this category cannot be initiated without submission of relevant information on the proposed experiment to NIH, review by the RAC Plant Working Group, and specific approval by NIH. Such experiments also require the approval of the IBC before initiation. Information on specific experiments which have been approved will be available in ORDA and will be listed in Appendix L-III when the Guidelines are republished.

Experiments which do not meet the specifications of Appendix L-II fall under Section III-A and require RAC review and NIH and IBC approval before initiation.

Appendix L-II.   Criteria Allowing Review by the RAC Plant Working Group Without the Requirement for Full RAC Review. Approval may be granted by ORDA in consultation with the Plant Working Group without the requirement for full RAC review (IBC review is also necessary) for growing plants containing recombinant DNA in the field under the following conditions:

Appendix L-II-A. The plant species is a cultivated crop of a genus that has no species known to be a noxious weed.

Appendix L-II-B. The introduced DNA consists of well-characterized genes containing no sequences harmful to humans, animals, or plants.

Appendix L-II-C. The vector consists of DNA: (i) From exempt host-vector systems (Appendix C); (ii) from plants of the same or closely related species; (iii) from nonpathogenic prokaryotes or nonpathogenic lower eukaryotic plants; (iv) from plant pathogens only if sequences resulting in production of disease symptoms have been deleted; or (v) chimeric vectors constructed from sequences defined in (i) to (iv) above. The DNA may be introduced by any suitable method. If sequences resulting in production of disease symptoms are retained for purposes of introducing the DNA into the plant, greenhouse-grown plants must be shown to be free of such sequences before such plants, derivatives, or seed from them can be used in field tests.

Appendix L-II-D. Plants are grown in controlled access fields under specified conditions appropriate for the plant under study and the geographical location. Such conditions should include provisions for using good cultural and pest control practices, for physical isolation from plants of the same species outside of the experimental plot in accordance with pollination characteristics of the species, and for further preventing plants containing recombinant DNA from becoming established in the environment. Review by the IBC should include an appraisal by scientists knowledgeable of the crop, its production practices, and the local geographical conditions. Procedures for assessing alterations in and the spread of organisms containing recombinant DNA must be developed. The results of the outlined tests must be submitted for review by the IBC. Copies must also be submitted to the Plant Working Group of the RAC.

Appendix L-III. Specific Approvals. As of publication of the revised Guidelines, no specific proposals have been approved. An updated list may be obtained from the Office of Recombinant DNA Activities, National Institutes of Health, Building 31, Room 3B10, Bethesda, Maryland 20205.

Dated: 15 November 1984.

JAMES B. WYNGAARDEN, M.D.
Director, National Institutes of Health

OMB's "Mandatory Information Requirements for Federal Assistance Program Announcements" (45 FR 39592) requires a statement concerning the official government programs contained in the *Catalog of Federal Domestic Assistance.* Normally NIH lists in its announcements the number and title of affected individual programs for the guidance of the public. Because the guidance in this notice covers not only virtually every NIH program but also essentially every federal research program in which DNA recombinant molecule techniques could be used, it has been determined to be not cost effective or in the public interest to attempt to list these programs. Such a list would likely require several additional pages. In addition, NIH could not be certain that every federal program would be included as many federal agencies, as well as private organizations, both national and international, have elected to follow the NIH Guidelines. In lieu of the individual program listing, NIH invites readers to direct questions to the information address above about whether individual programs listed in the *Catalog of Federal Domestic Assistance are affected.*

Annex II

UNITED STATES FOOD AND DRUG ADMINISTRATION GOOD MANUFACTURING PRACTICE REGULATIONS[a]

Part 210—Current good manufacturing practices in manufacturing, processing, packing, or holding of drugs: general

Sec.

210.1   Status of current good manufacturing practice regulations.
210.2   Applicability of current good manufacturing practice regulations.
210.3   Definitions.

Authority: Secs. 501, 701, 52 Stat. 1049-1050 as amended, 1055-1056 as amended (21 U.S.C. 351, 371).

Source: 43 FR 4:076. Sept. 29, 1978, unless otherwise noted.

§210.1   *Status of current good manufacturing practice regulations.*

(a)   The regulations set forth in this part and in Parts 211 through 229 of this chapter contain the minimum current good manufacturing practice for methods to be used in, and the facilities or controls to be used for, the manufacture, processing, packing, or holding of a drug to assure that such drug meets the requirements of the act as to safety, and has the identity and strength and meets the quality and purity characteristics that it purports or is represented to possess.

(b)   The failure to comply with any regulation set forth in this part and in Parts 211 through 229 of this chapter in the manufacture, processing, packing, or holding of a drug shall render such drug to be adulterated under section 501(a)(2)(B) of the act and such drug, as well as the person who is responsible for the failure to comply, shall be subject to regulatory action.

§210.2   *Applicability of current good manufacturing practice regulations.*

(a)   The regulations in this part and in Parts 211 through 229 of this chapter as they may pertain to a drug and in Parts 600 through 680 of this chapter as they may pertain to a biological product for human use, shall be considered to supplement, not supersede, each other, unless the regulations explicitly provide otherwise. In the event that it is impossible to comply with all applicable regulations in these parts, the regulations specifically applicable to the drug in question shall supersede the more general.

(b)   If a person engages in only some operations subject to the regulations in this part and in Parts 211 through 229 and Parts 600 through 680 of this chapter, and not in others, that person need only comply with those regulations applicable to the operations in which he or she is engaged.

---

[a]United States of America, 21 C.F.R. (1985).

§210.3 *Definitions.*

(a) The definitions and interpretations contained in section 201 of the act shall be applicable to such terms when used in this part and in Parts 211 through 229 of this chapter.

(b) The following definitions of terms apply to this part and to Parts 211 through 229 of this chapter.

(1) "Act" means the Federal Food, Drug, and Cosmetic Act, as amended (21 U.S.C. d301 et seq.).

(2) "Batch" means a specific quantity of a drug or other material that is intended to have uniform character and quality, within specified limits, and is produced according to a single manufacturing order during the same cycle of manufacture.

(3) "Component" means any ingredient intended for use in the manufacture of a drug product, including those that may not appear in such drug product.

(4) "Drug product" means a finished dosage form, for example, tablet, capsule, solution, etc., that contains an active drug ingredient generally, but not necessarily, in association with inactive ingredients. The term also includes a finished dosage form that does not contain an active ingredient but is intended to be used as a placebo.

(5) "Fiber" means any particulate contaminant with a length at least three times greater than its width.

(6) "Non-fiber-releasing filter" means any filter, which after any appropriate pretreatment such as washing or flushing, will not release fibers into the component or drug product that is being filtered. All filters composed of asbestos are deemed to be fiber-releasing filters.

(7) "Active ingredient" means any component that is intended to furnish pharmacological activity or other direct effect in the diagnosis, cure, mitigation, treatment, or prevention of disease, or to affect the structure or any function of the body of man or other animals. The term includes those components that may undergo chemical change in the manufacture of the drug product and be present in the drug product in a modified form intended to furnish the specified activity or effect.

(8) "Inactive ingredient" means any component other than an "active ingredient."

(9) "In-process material" means any material fabricated, compounded, blended, or derived by chemical reaction that is produced for, and used in, the preparation of the drug product.

(10) "Lot" means a batch, or a specific identified portion of a batch, having uniform character and quality within specified limits: or, in the case of a drug product produced by continuous process, it is a specific identified amount produced in a unit of time or quantity in a manner that assures its having uniform character and quality within specified limits.

(11) "Lot number, control number, or batch number" means any distinctive combination of letters, numbers, or symbols, or any combination of them, from which the complete history of the manufacture, processing, packing, holding, and distribution of a batch or lot of drug product or other material can be determined.

(12) "Manufacture, processing, packing, or holding of a drug product" includes packaging and labeling operations, testing, and quality control of drug products.

(13) "Medicated feed" means any "complete feed," "feed supplement," or "feed concentrate" as defined in §558.3 of this chapter and is a feed that contains one or more drugs as defined in section 201(g) of the act. Medicated feeds are subject to Part 225 of this chapter.

(14) "Medicated premix" means a substance that meets the definition in §558.3 of this chapter for a "feed premix," except that it contains one or more drugs as defined in section 201(g) of the act and is intended for manufacturing use in the production of a medicated feed. Medicated premixes are subject to Part 226 of this chapter.

(15) "Quality control unit" means any person or organizational element designated by the firm to be responsible for the duties relating to quality control.

(16) "Strength" means:

    (i) The concentration of the drug substance (for example, weight/weight, weight/volume, or unit dose/volume basis), and/or

    (ii) The potency, that is, the therapeutic activity of the drug product as indicated by appropriate laboratory tests or by adequately developed and controlled clinical data (expressed, for example, in terms of units by reference to a standard).

(17) "Theoretical yield" means the quantity that would be produced at any appropriate phase of manufacture, processing, or packing of a particular drug product, based upon the quantity of components to be used, in the absence of any loss or error in actual production.

Food and Drug Administration HHS

(18) "Actual yield" means the quantity that is actually produced at any appropriate phase of manufacture, processing, or packing of a particular drug product.

(19) "Percentage of theoretical yield" means the ratio of the actual yield (at any appropriate phase of manufacture, processing, or packing of a particular drug product) to the theoretical yield (at the same phase), stated as a percentage.

(20) "Acceptance criteria" means the product specifications and acceptance/rejection criteria, such as acceptable quality level and unacceptable quality level, with an associated sampling plan, that are necessary for making a decision to accept or reject a lot or batch (or any other convenient subgroups of manufactured units).

(21) "Representative sample" means a sample that consists of a number of units that are drawn based on rational criteria such as random sampling and intended to assure that the sample accurately portrays the material being sampled.

Part 211—Current good manufacturing practice for finished pharmaceuticals

Subpart A–General Provisions

Subpart B–Organization and Personnel

Subpart C–Buildings and Facilities

211.42   Design and construction features.
211.44   Lighting.
211.46   Ventilation, air filtration, air heating and cooling.
211.48   Plumbing.
211.50   Sewage and refuse.
211.52   Washing and toilet facilities.
211.56   Sanitation.
211.58   Maintenance.

Subpart D–Equipment

211.63   Equipment design, size, and location.
211.65   Equipment construction.
211.67   Equipment cleaning and maintenance.
211.68   Automatic, mechanical, and electronic equipment
211.72   Filters.

Subpart E–Control of Components and Drug Product Containers and Closures

211.80   General requirements.
211.82   Receipt and storage of untested components, drug product containers, and closures.
211.84   Testing and approval or rejection of components, drug product containers, and closures.
211.86   Use of approved components, drug product containers, and closures.
211.87   Retesting of approved components, drug product containers, and closures.
211.89   Rejected components, drug product containers, and closures.
211.94   Drug product containers and closures.

Subpart F–Production and Process Controls

211.100   Written procedures; deviations.
211.101   Charge-in of components.
211.103   Calculation of yield.
211.105   Equipment identification.
211.110   Sampling and testing of in-process materials and drug products.
211.111   Time limitations on production.
211.113   Control of microbiological contamination.
211.115   Reprocessing.

Subpart G–Packaging and Labeling Control

211.122   Materials examination and usage criteria.
211.125   Labeling issuance.
211.130   Packaging and labeling operations.
211.132   Tamper-resistant packaging requirements for over-the-counter human drug products.
211.134   Drug product inspection.
211.137   Expiration dating.

Subpart H–Holding and Distribution

211.142   Warehousing procedures.
211.150   Distribution procedures.

108

Subpart I–Laboratory Controls

211.160   General requirements.
211.165   Testing and release for distribution.
211.166   Stability testing.
211.167   Special testing requirements.
211.170   Reserve samples.
211.173   Laboratory animals
211.176   Penicillin contamination.

Subpart J–Records and Reports

211.180   General requirements.
211.182   Equipment cleaning and use log.
211.184   Component, drug product container, closure, and labeling records.
211.186   Master production and control records.
211.188   Batch production and control records.
211.192   Production record review.
211.194   Laboratory records.
211.196   Distribution records.
211.198   Complaint files.

Subpart K–Returned and Salvaged Drug Products

211.204   Returned drug products.
211.208   Drug product salvaging.

Authority: Secs. 501, 701, 52 Stat. 1049-1050 as amended, 1055-1056 as amended (21 U.S.C. 351, 371).

Source: 43 FR 45077, Sept. 29, 1978, unless otherwise noted.

Subpart A   General Provisions

§211.1   *Scope.*

(a)   The regulations in this part contain the minimum current good manufacturing practice for preparation of drug products for administration to humans or animals.

(b)   The current good manufacturing practice regulations in this chapter, as they pertain to drug products, and in Parts 600 through 680 of this chapter, as they pertain to biological products for human use, shall be considered to supplement, not supersede, the regulations in this part unless the regulations explicitly provide otherwise. In the event it is impossible to comply with applicable regulations both in this part and in other parts of this chapter or in Parts 600 through 680 of this chapter, the regulation specifically applicable to the drug product in question shall supersede the regulation in this part.

(c)   Pending consideration of a proposed exemption, published in the FEDERAL REGISTER of September 29, 1978, the requirements in this part shall not be enforced for OTC drug products if the products and all their ingredients are ordinarily marketed and consumed as human foods, and which products may also fall within the legal definition of drugs by virtue of their intended use. Therefore, until further notice, regulations under Part 110 of this chapter, and where applicable, Parts 113 to 129 of this chapter, shall be applied in determining whether these OTC drug products that are also foods are manufactured, processed, packed, or held under current good manufacturing practice.

§211.3   *Definitions.*

The definitions set forth in paragraph 210.3 of this chapter apply in this part.

Subpart B.   Organization and Personnel

§211.22   *Responsibilities of quality control unit.*

(a)   There shall be a quality control unit that shall have the responsibility and authority to approve or reject all components, drug product containers, closures, in-process materials, packaging material, labeling, and drug products, and the authority to review production records to assure that no errors have occurred or, if errors have occurred, that they have been fully investigated. The quality control unit shall be responsible for approving or rejecting drug products manufactured, processed, packed, or held under contract by another company.

(b)   Adequate laboratory facilities for the testing and approval (or rejection) of components, drug product containers, closures, packaging materials, in-process materials, and drug products shall be available to the quality control unit.

(c)   The quality control unit shall have the responsibility for approving or rejecting all procedures or specifications impacting on the identity, strength, quality, and purity of the drug product.

(d)   The responsibilities and procedures applicable to the quality control unit shall be in writing: such written procedures shall be followed.

§211.25   *Personnel qualifications.*

(a)   Each person engaged in the manufacture, processing, packing, or holding of a drug product shall have education, training, and experience, or any combination thereof, to enable that person to perform the assigned functions. Training shall be in the particular operations that the employee performs and in current good manufacturing practice (including the current good manufacturing practice regulations in this chapter and written procedures required by these regulations) as they relate to the employee's functions. Training in current good manufacturing practice shall be conducted by qualified individuals on a continuing basis and with sufficient frequency to assure that employees remain familiar with CGMP requirements applicable to them.

(b)   Each person responsible for supervising the manufacture, processing, packing, or holding of a drug product shall have the education, training, and experience, or any combination thereof, to perform assigned functions in such a manner as to provide assurance that the drug product has the safety, identity, strength, quality, and purity that it purports or is represented to possess.

(c)   There shall be an adequate number of qualified personnel to perform and supervise the manufacture, processing, packing, or holding of each drug product.

§211.28   *Personnel responsibilities.*

(a)   Personnel engaged in the manufacture, processing, packing, or holding of a drug product shall wear clean clothing appropriate for the duties they perform. Protective apparel, such as head, face, hand, and arm coverings, shall be worn as necessary to protect drug products from contamination.

(b)   Personnel shall practice good sanitation and health habits.

(c)   Only personnel authorized by supervisory personnel shall enter those areas of the buildings and facilities designated as limited-access areas.

(d)   Any person shown at any time (either by medical examination or supervisory observation) to have an apparent illness or open lesions that may adversely affect the safety or quality of drug products shall be excluded from direct contact with components, drug product containers, closures, in-process materials, and drug products until the condition is corrected or determined by competent medical personnel not to jeopardize the safety or quality of drug products. All personnel shall be instructed to report to supervisory personnel any health conditions that may have an adverse effect on drug products.

§211.34 *Consultants.*

Consultants advising on the manufacture, processing, packing, or holding of drug products shall have sufficient education, training, and experience, or any combination thereof, to advise on the subject for which they are retained. Records shall be maintained stating the name, address, and qualifications of any consultants and the type of service they provide.

Subpart C. Buildings and Facilities

§211.42 *Design and construction features.*

(a) Any building or buildings used in the manufacture, processing, packing, or holding of a drug product shall be of suitable size, construction and location to facilitate cleaning, maintenance, and proper operations.

(b) Any such building shall have adequate space for the orderly placement of equipment and materials to prevent mixups between different components, drug product containers, closures, labeling, in-process materials, or drug products, and to prevent contamination. The flow of components, drug product containers, closures, labeling, in-process materials, and drug products through the building or buildings shall be designed to prevent contamination.

(c) Operations shall be performed within specifically defined areas of adequate size. There shall be separate or defined areas for the firm's operations to prevent contamination or mixups as follows:

(1) Receipt, identification, storage, and withholding from use of components, drug product containers, closures, and labeling, pending the appropriate sampling, testing, or examination by the quality control unit before release for manufacturing or packaging;

(2) Holding rejected components, drug product containers, closures, and labeling before disposition;

(3) Storage of released components, drug product containers, closures, and labeling;

(4) Storage of in-process materials;

(5) Manufacturing and processing operations;

(6) Packaging and labeling operations;

(7) Quarantine storage before release of drug products;

(8) Storage of drug products after release;

(9) Control and laboratory operations;

(10) Aseptic processing, which includes as appropriate.

    (i) Floors, walls, and ceilings of smooth, hard surfaces that are easily cleanable;

    (ii) Temperature and humidity controls;

    (iii) An air supply filtered through high-efficiency particulate air filters under positive pressure, regardless of whether flow is laminar or nonlaminar;

    (iv) A system for monitoring environmental conditions;

    (v) A system for cleaning and disinfecting the room and equipment to produce aseptic conditions;

(vi) A system for maintaining any equipment used to control the aseptic conditions.

(d) Operations relating to the manufacture, processing, and packing of penicillin shall be performed in facilities separate from those used for other drug products for human use.

§211.44 Lighting

Adequate lighting shall be provided in all areas.

§211.46 Ventilation, air filtration, air heating and cooling.

(a) Adequate ventilation shall be provided.

(b) Equipment for adequate control over air pressure, microorganisms, dust, humidity, and temperature shall be provided when appropriate for the manufacture, processing, packing, or holding of a drug product.

(c) Air filtration systems, including prefilters and particulate matter air filters, shall be used when appropriate on air supplies to production areas. If air is recirculated to production areas, measures shall be taken to control recirculation of dust from production. In areas where air contamination occurs during production, there shall be adequate exhaust systems or other systems adequate to control contaminants.

(d) Air-handling systems for the manufacture, processing, and packing of penicillin shall be completely separate from those for other drug products for human use.

§211.48 Plumbing.

(a) Potable water shall be supplied under continuous positive pressure in a plumbing system free of defects that could contribute contamination to any drug product. Potable water shall meet the standards prescribed in the Environmental Protection Agency's Primary Drinking Water Regulations set forth in 40 CFR Part 141. Water not meeting such standards shall not be permitted in the potable water system.

(b) Drains shall be of adequate size and, where connected directly to a sewer, shall be provided with an air break or other mechanical device to prevent back-siphonage.

43 FR 45077, Sept. 29, 1978, as amended at 48 FR 11426, Mar. 18, 1983

§211.50 Sewage and refuse.

Sewage, trash, and other refuse in and from the building and immediate premises shall be disposed of in a safe and sanitary manner.

§211.52 Washing and toilet facilities.

Adequate washing facilities shall be provided, including hot and cold water, soap or detergent, air driers or single-service towels, and clean toilet facilities easily accessible to working areas.

§211.56 Sanitation.

(a) Any building used in the manufacture, processing, packing, or holding of a drug product shall be maintained in a clean and sanitary condition. Any such building shall be free of infestation by rodents, birds, insects and other vermin (other than laboratory animals). Trash and organic waste matter shall be held and disposed of in a timely and sanitary manner.

(b) There shall be written procedures assigning responsibility for sanitation and describing in sufficient detail the cleaning schedules, methods, equipment, and materials to be used in cleaning the buildings and facilities: such written procedures shall be followed.

(c) There shall be written procedures for use of suitable rodenticides, insecticides, fungicides, fumigating agents, and cleaning and sanitizing agents. Such written procedures shall be designed to prevent the contamination of equipment, components, drug product containers, closures, packaging, labeling materials, or drug products, and shall be followed. Rodenticides, insecticides, and fungicides shall not be used unless registered and used in accordance with the Federal Insecticide, Fungicide, and Rodenticide Act (7 U.S.C. 135).

(d) Sanitation procedures shall apply to work performed by contractors or temporary employees as well as work performed by full-time employees during the ordinary course of operations.

§211.58 *Maintenance.*

Any building used in the manufacture, processing, packing, or holding of a drug product shall be maintained in a good state of repair.

Subpart D. Equipment

§211.63 *Equipment design, size, and location.*

Equipment used in the manufacture, processing, packing, or holding of a drug product shall be of appropriate design, adequate size, and suitably located to facilitate operations for its intended use and for its cleaning and maintenance.

§211.65 *Equipment construction.*

(a) Equipment shall be constructed so that surfaces that contact components, in-process materials, or drug products shall not be reactive, additive, or absorptive so as to alter the safety, identity, strength, quality, or purity of the drug product beyond the official or other established requirements.

(b) Any substances required for operation, such as lubricants or coolants, shall not come into contact with components, drug product containers, closures, in-process materials, or drug products so as to alter the safety, identity, strength, quality, or purity of the drug product beyond the official or other established requirements.

§211.67 *Equipment cleaning and maintenance.*

(a) Equipment and utensils shall be cleaned, maintained, and sanitized at appropriate intervals to prevent malfunctions or contamination that would alter the safety, identity, strength, quality, or purity of the drug product beyond the official or other established requirements.

(b) Written procedures shall be established and followed for cleaning and maintenance of equipment, including utensils, used in the manufacture, processing, packing, or holding of a drug product. These procedures shall include, but are not necessarily limited to, the following:

(1) Assignment of responsibility for cleaning and maintaining equipment;

(2) Maintenance and cleaning schedules, including, where appropriate, sanitizing schedules;

(3) A description in sufficient detail of the methods, equipment, and materials used in cleaning and maintenance operations, and the methods of disassembling and reassembling equipment as necessary to assure proper cleaning and maintenance;

(4) Removal or obliteration of previous batch identification;

(5) Protection of clean equipment from contamination prior to use;

(6) Inspection of equipment for cleanliness immediately before use.

(c) Records shall be kept of maintenance, cleaning, sanitizing, and inspection as specified in §211.180 and §211.182.

§211.68 *Automatic, mechanical, and electronic equipment.*

(a) Automatic, mechanical, or electronic equipment or other types of equipment, including computers, or related systems that will perform a function satisfactorily, may be used in the manufacture, processing, packing, and holding of a drug product. If such equipment is so used, it shall be routinely calibrated, inspected, or checked according to a written program designed to assure proper performance. Written records of those calibration checks and inspections shall be maintained.

(b) Appropriate controls shall be exercised over computer or related systems to assure that changes in master production and control records or other records are instituted only by authorized personnel. Input to and output from the computer or related system of formulas or other records or data shall be checked for accuracy. A backup file of data entered into the computer or related system shall be maintained except where certain data, such as calculations performed in connection with laboratory analysis, are eliminated by computerization or other automated processes. In such instances a written record of the program shall be maintained along with appropriate validation data. Hard copy or alternative systems, such as duplicates, tapes, or microfilm, designed to assure that backup data are exact and complete and that it is secure from alteration, inadvertent erasures, or loss shall be maintained.

§211.72 *Filters.*

Filters for liquid filtration used in the manufacture, processing, or packing of injectable drug products intended for human use shall not release fibers into such products. Fiber-releasing filters may not be used in the manufacture, processing, or packing of these injectable drug products unless it is not possible to manufacture such drug products without the use of such filters. If use of a fiber-releasing filter is necessary, an additional non-fiber-releasing filter of 0.22 micron maximum mean porosity (0.45 micron if the manufacturing conditions so dictate) shall subsequently be used to reduce the content of particles in the injectable drug product. Use of an asbestos-containing filter, with or without subsequent use of a specific non-fiber-releasing filter, is permissible only upon submission of proof to the appropriate bureau of the Food and Drug Administration that use of a non-fiber-releasing filter will, or is likely to, compromise the safety or effectiveness of the injectable drug product.

Subpart E. *Control of Components and Drug Product Containers and Closures*

§211.80 *General requirements.*

(a) There shall be written procedures describing in sufficient detail the receipt, identification, storage, handling, sampling, testing, and approval or rejection of components and drug product containers and closures; such written procedures shall be followed.

(b) Components and drug product containers and closures shall at all times be handled and stored in a manner to prevent contamination.

(c) Bagged or boxed components of drug product containers, or closures shall be stored off the floor and suitably spaced to permit cleaning and inspection.

(d) Each container or grouping of containers for components or drug product containers, or closures shall be identified with a distinctive code for each lot in each shipment received. This code shall be used in recording the disposition of each lot. Each lot shall be appropriately identified as to its status (i.e., quarantined, approved, or rejected).

§211.82 *Receipt and storage of untested components, drug product containers, and closures.*

(a) Upon receipt and before acceptance, each container or grouping of containers of components, drug product containers, and closures shall be examined visually for appropriate labeling as to contents, container damage or broken seals, and contamination.

(b) Components, drug product containers, and closures shall be stored under quarantine until they have been tested or examined, as appropriate, and released. Storage within the area shall conform to the requirements of §211.80.

§211.84 *Testing and approval or rejection of components, drug product containers, and closures.*

(a) Each lot of components, drug product containers, and closures shall be withheld from use until the lot has been sampled, tested, or examined, as appropriate, and released for use by the quality control unit.

(b) Representative samples of each shipment of each lot shall be collected for testing or examination. The number of containers to be sampled, and the amount of material to be taken from each container, shall be based upon appropriate criteria such as statistical criteria for component variability, confidence levels, and degree of precision desired, the past quality history of the supplier, and the quantity needed for analysis and reserve where required by §211.170.

(c) Samples shall be collected in accordance with the following procedures:

(1) The containers of components selected shall be cleaned where necessary, by appropriate means.

(2) The containers shall be opened, sampled, and resealed in a manner designed to prevent contamination of their contents and contamination of other components, drug product containers, or closures.

(3) Sterile equipment and aseptic sampling techniques shall be used when necessary.

(4) If it is necessary to sample a component from the top, middle, and bottom of its container, such sample subdivisions shall not be composited for testing.

(5) Sample containers shall be identified so that the following information can be determined: name of the material sampled, the lot number, the container from which the sample was taken, the data on which the sample was taken, and the name of the person who collected the sample.

(6) Containers from which samples have been taken shall be marked to show that samples have been removed from them.

(d) Samples shall be examined and tested as follows:

(1) At least one test shall be conducted to verify the identity of each component of a drug product. Specific identity tests, if they exist, shall be used.

(2) Each component shall be tested for conformity with all appropriate written specifications for purity, strength, and quality. In lieu of such testing by the manufacturer, a report of analysis may be accepted from the supplier of a component, provided that at least one specific identity test is conducted on such component by the manufacturer, and provided that the manufacturer establishes the reliability of the supplier's analyses through appropriate validation of the supplier's test results at appropriate intervals.

(3) Containers and closures shall be tested for conformance with all appropriate written procedures. In lieu of such testing by the manufacturer, a certificate of testing may be accepted from the supplier, provided that at least a visual identification is conducted on such containers/closures by the manufacturer and provided that the manufacturer establishes the reliability of the supplier's test results through appropriate validation of the supplier's test results at appropriate intervals.

(4) When appropriate, components shall be microscopically examined.

(5) Each lot of a component, drug product container, or closure that is liable to contamination with filth, insect infestation, or other extraneous adulterant shall be examined against established specifications for such contamination.

(6) Each lot of a component, drug product container, or closure that is liable to microbiological contamination that is objectionable in view of its intended use shall be subjected to microbiological tests before use.

(e) Any lot of components, drug product containers, or closures that meets the appropriate written specifications of identity, strength, quality, and purity and related tests under paragraph *(d)* of this section may be approved and released for use. Any lot of such material that does not meet such specifications shall be rejected.

§211.86 *Use of approved components, drug product containers, and closures.*

Components, drug product containers, and closures approved for use shall be rotated so that the oldest approved stock is used first. Deviation from this requirement is permitted if such deviation is temporary and appropriate.

§211.87 *Retesting of approved components, drug product containers, and closures.*

Components, drug product containers, and closures shall be retested or reexamined, as appropriate, for identity, strength, quality, and purity and approved or rejected by the quality control unit in accordance with §211.84 as necessary, e.g., after storage for long periods or after exposure to air, heat or other conditions that might adversely affect the component, drug product container, or closure.

§211.89 *Rejected components, drug product containers, and closures.*

Rejected components, drug product containers, and closures shall be identified and controlled under a quarantine system designed to prevent their use in manufacturing or processing operations for which they are unsuitable.

§211.94 *Drug product containers and closures.*

(a) Drug product containers and closures shall not be reactive, additive, or absorptive so as to alter the safety, identity, strength, quality, or purity of the drug beyond the official or established requirements.

(b) Container closure systems shall provide adequate protection against foreseeable external factors in storage and use that can cause deterioration or contamination of the drug product.

116

(c) Drug product containers and closures shall be clean and, where indicated by the nature of the drug, sterilized and processed to remove pyrogenic properties to assure that they are suitable for their intended use.

(d) Standards or specifications, methods of testing, and, where indicated, methods of cleaning, sterilizing, and processing to remove pyrogenic properties shall be written and followed for drug product containers and closures.

Subpart F. Production and Process Controls

§211.100 *Written procedures; deviations.*

(a) There shall be written procedures for production and process control designed to assure that the drug products have the identity, strength, quality, and purity they purport or are represented to possess. Such procedures shall include all requirements in this subpart. These written procedures, including any changes, shall be drafted, reviewed, and approved by the appropriate organizational units and reviewed and approved by the quality control unit.

(b) Written production and process control procedures shall be followed in the execution of the various production and process control functions and shall be documented at the time of performance. Any deviation from the written procedures shall be recorded and justified.

§211.101 *Charge-in of components.*

Written production and control procedures shall include the following, which are designed to assure that the drug products produced have the identity, strength, quality, and purity they purport or are represented to possess:

(a) The batch shall be formulated with the intent to provide not less than 100 per cent of the labeled or established amount of active ingredient.

(b) Components for drug product manufacturing shall be weighed, measured, or subdivided as appropriate. If a component is removed from the original container to another, the new container shall be identified with the following information:

(1) Component name or item code;

(2) Receiving or control number;

(3) Weight or measure in new container;

(4) Batch for which component was dispensed, including its product name, strength, and lot number.

(c) Weighing, measuring, or subdividing operations for components shall be adequately supervised. Each container of component dispensed to manufacturing shall be examined by a second person to assure that:

(1) The component was released by the quality control unit;

(2) The weight or measure is correct as stated in the batch production records;

(3) The containers are properly identified.

(d) Each component shall be added to the batch by one person and verified by a second person.

§211.103 *Calculation of yield.*

Actual yields and percentages of theoretical yield shall be determined at the conclusion of each appropriate phase of manufacturing, processing, packaging, or holding of the drug product. Such calculations shall be performed by one person and independently verified by a second person.

117

§211.105 Equipment identification.

(a) All compounding and storage containers, processing lines, and major equipment used during the production of a batch of a drug product shall be properly identified at all times to indicate their contents and, when necessary, the phase of processing of the batch.

(b) Major equipment shall be identified by a distinctive identification number or code that shall be recorded in the batch production record to show the specific equipment used in the manufacture of each batch of a drug product. In cases where only one of a particular type of equipment exists in a manufacturing facility, the name of the equipment may be used in lieu of a distinctive identification number or code.

§211.110 Sampling and testing of in-process materials and drug products.

(a) To assure batch uniformity and integrity of drug products, written procedures shall be established and followed that describe the in-process controls, and tests, or examinations to be conducted on appropriate samples of in-process materials of each batch. Such control procedures shall be established to monitor the output and to validate the performance of those manufacturing processes that may be responsible for causing variability in the characteristics of in-process material and the drug product. Such control procedures shall include, but are not limited to, the following, where appropriate:

(1) Tablet or capsule weight variation;

(2) Disintegration time;

(3) Adequacy of mixing to assure uniformity and homogeneity;

(4) Dissolution time and rate;

(5) Clarity, completeness, or pH of solutions.

(b) Valid in-process specifications for such characteristics shall be consistent with drug product final specifications and shall be derived from previous acceptable process average and process variability estimates where possible and determined by the application of suitable statistical procedures where appropriate. Examination and testing of samples shall assure that the drug product and in-process material conform to specifications.

(c) In-process materials shall be tested for identity, strength, quality, and purity as appropriate, and approved or rejected by the quality control unit, during the production process, e.g., at commencement or completion of significant phases or after storage for long periods.

(d) Rejected in-process materials shall be identified and controlled under a quarantine system designed to prevent their use in manufacturing or processing operations for which they are unsuitable.

§211.111 Time limitations on production.

When appropriate, time limits for the completion of each phase of production shall be established to assure the quality of the drug product. Deviation from established time limits may be acceptable if such deviation does not compromise the quality of the drug product. Such deviation shall be justified and documented.

§211.113 Control of microbiological contamination.

(a) Appropriate written procedures, designed to prevent objectionable microorganisms in drug products not required to be sterile, shall be established and followed.

118

(b) Appropriate written procedures, designed to prevent microbiological contamination of drug products purporting to be sterile, shall be established and followed. Such procedures shall include validation of any sterilization process.

§211.115 Reprocessing.

(a) Written procedures shall be established and followed prescribing a system for reprocessing batches that do not conform to standards or specifications and the steps to be taken to insure that the reprocessed batches will conform with all established standards, specifications, and characteristics.

(b) Reprocessing shall not be performed without the review and approval of the quality control unit.

Subpart G. Packaging and Labeling Control

§211.122 Materials examination and usage criteria.

(a) There shall be written procedures describing in sufficient detail the receipt, identification, storage, handling, sampling, examination, and/or testing of labeling and packaging materials; such written procedures shall be followed. Labeling and packaging materials shall be representatively sampled, and examined or tested upon receipt and before use in packaging or labeling of a drug product.

(b) Any labeling or packaging materials meeting appropriate written specifications may be approved and released for use. Any labeling or packaging materials that do not meet such specifications shall be rejected to prevent their use in operations for which they are unsuitable.

(c) Records shall be maintained for each shipment received of each different labeling and packaging material indicating receipt, examination or testing, and whether accepted or rejected.

(d) Labels and other labeling materials for each different drug product, strength, dosage form, or quantity of contents shall be stored separately with suitable identification. Access to the storage area shall be limited to authorized personnel.

(e) Obsolete and outdated labels, labeling, and other packaging materials shall be destroyed.

(f) Gang printing of labeling to be used for different drug products or different strengths of the same drug product (or labeling of the same size and identical or similar format and/or color schemes) shall be minimized. If gang printing is employed, packaging and labeling operations shall provide for special control procedures, taking into consideration sheet layout, stacking, cutting, and handling during and after printing.

(g) Printing devices on, or associated with, manufacturing lines used to imprint labeling upon the drug product unit label or case shall be monitored to assure that all imprinting conforms to the print specified in the batch production record.

§211.125 Labeling issuance.

(a) Strict control shall be exercised over labeling issued for use in drug product labeling operations.

(b) Labeling materials issued for a batch shall be carefully examined for identity and conformity to the labeling specified in the master or batch production records.

(c) Procedures shall be utilized to reconcile the quantities of labeling issued, used, and returned, and shall require evaluation of discrepancies found between the quantity of drug product finished and the quantity of labeling issued when such discrepancies are outside narrow preset limits based on historical operating data. Such discrepancies shall be investigated in accordance with §211.192.

(d) All excess labeling bearing lot or control numbers shall be destroyed.

(e) Returned labeling shall be maintained and stored in a manner to prevent mixups and provide proper identification.

(f) Procedures shall be written describing in sufficient detail the control procedures employed for the issuance of labeling: such written procedures shall be followed.

§211.130 *Packaging and labeling operations.*

There shall be written procedures designed to assure that correct labels, labeling, and packaging materials are used for drug products: such written procedures shall be followed. These procedures shall incorporate the following features:

(a) Prevention of mixups and cross-contamination by physical or spatial separation from operations on other drug products.

(b) Identification of the drug product with a lot or control number that permits determination of the history of the manufacture and control of the batch.

(c) Examination of packaging and labeling materials for suitability and correctness before packaging operations, and documentation of such examination in the batch production record.

(d) Inspection of the packaging and labeling facilities immediately before use to assure that all drug products have been removed from previous operations. Inspection shall also be made to assure that packaging and labeling materials not suitable for subsequent operations have been removed. Results of inspection shall be documented in the batch production records.

§211.132 *Tamper-resistant packaging requirements for over-the-counter human drug products.*

(a) *General.* Because most over-the-counter (OTC) human drug products are not now packaged in tamper-resistant retail packages, there is the opportunity for the malicious adulteration of OTC drug products with health risks to individuals who unknowingly purchase adulterated products and with loss of consumer confidence in the security of OTC drug product packages. The Food and Drug Administration has the authority and responsibility under the Federal Food, Drug, and Cosmetic Act (the act) to establish a uniform national requirement for tamper-resistant packaging of OTC drug products that will improve the security of OTC drug packing and help assure the safety and effectiveness of OTC drug products. An OTC drug product (except a dermatological, dentifrice, insulin, or lozenge product) for retail sale that is not packaged in a tamper-resistant package or that is not properly labeled under this section is adulterated under section 501 of the act or misbranded under section 502 of the act, or both.

(b) *Requirement for tamper-resistant package.* Each manufacturer and packer who packages an OTC drug product (except a dermatological, dentifrice, insulin or lozenge product) for retail sale, shall package the product in a tamper-resistant package, if this product is accessible to the public while held for sale. A tamper-resistant package is one having an indicator or barrier to entry which, if breached or missing, can reasonably be expected to provide visible evidence to consumers that tampering has occurred. To reduce the likelihood of substitution of a tamper-resistant feature after tampering, the indicator or barrier to entry is required to be distinctive by design (e.g., an aerosol product container) or by the use of an identifying characteristic (e.g., a pattern, name, registered trademark, logo, or picture). For purposes of this section, the term "distinctive by design" means the packaging cannot be duplicated with commonly available materials or through commonly available processes. For purposes of this section, the term "aerosol product" means a product which depends upon the power of a liquified or compressed gas to expel the contents from the container. A tamper-resistant package may involve an immediate-container and closure system or secondary-

120

container of carton system or any combination of systems intended to provide a visual indication of package integrity. The tamper-resistant feature shall be designed to and shall remain intact when handled in a reasonable manner during manufacture, distribution, and retail display.

(c) Labeling. Each retail package of an OTC drug product covered by this section, except ammonia inhalant in crushable glass ampules, aerosol products as defined in paragraph *(b)* of this section, or containers of compressed medical oxygen, is required to bear a statement that is prominently placed so that consumers are alerted to the specific tamper-resistant feature of the package. The labeling statement is also required to be so placed that it will be unaffected if the tamper-resistant feature of the package is breached or missing. If the tamper-resistant feature chosen to meet the requirement in paragraph *(b)* of this section is one that uses an identifying characteristic, that characteristic is required to be referred to in the labeling statement. For example, the labeling statement on a bottle with a shrink band could say "For your protection, this bottle has an imprinted seal around the neck".

(d) Requests for exemptions from packaging and labeling requirements. A manufacturer or packer may request an exemption from the packaging and labeling requirements of this section. A request for an exemption is required to be submitted in the form of a citizen petition under §10.30 of this chapter and should be clearly identified on the envelope as a "Request for Exemption from Tamper-resistant Rule". The petition is required to contain the following:

(1)   The name of the drug product or, if the petition seeks an exemption for a drug class, the name of the drug class, and a list of products within that class.

(2)   The reasons that the drug product's compliance with the tamper-resistant packaging or labeling requirements of this section is unnecessary or cannot be achieved.

(3)   A description of alternative steps that are available, or that the petitioner has already taken, to reduce the likelihood that the product or drug class will be the subject of malicious adulteration.

(4)   Other information justifying an exemption.

This information collection requirement has been approved by the Office of Management and Budget under number 0910-0149.

(e)   OTC drug products subject to approved new drug applications. Holders of approved new drug applications for OTC drug products are required under §314.8 *(a)* (4)(vi), (5)(xi), or *(d)*(5) of this chapter to provide for changes in packaging, and under §314.8*(a)*(5)(xii) to provide for changes in labeling to comply with the requirements of this section.

(f)   Poison Prevention Packaging Act of 1970. This section does not affect any requirements for "special packaging" as defined under §310.3(1) of this chapter and required under the Poison Prevention Packaging Act of 1970.

(g)   Effective date. OTC drug products, except dermatological, dentifrice, insulin, and lozenge products, are required to comply with the requirements of this section on the dates listed below except to the extent that a product's manufacturer or packer has obtained an exemption from a packaging or labeling requirement.

(1)   *Initial effective date for packaging requirements.* (i) The packaging requirements in paragraph *(b)* of this section is effective on February 7, 1983 for each affected OTC drug product (except oral and vaginal tablets, vaginal and rectal suppositories, and one-piece soft gelatin capsules) packaged for retail sale on or after that date, except for the requirement in paragraph *(b)* of this section for a distinctive indicator or barrier to entry.

(ii)   The packaging requirement in paragraph *(b)* of this section is effective on May 5, 1983 for each OTC drug product that is an oral or vaginal tablet, a vaginal or rectal suppository, or one-piece soft gelatin capsules packaged for retail sale on or after that date.

(2)   *Initial effective date for labeling requirements.* The requirement in paragraph *(b)* of this section that the indicator or barrier to entry be distinctive by design and the requirement in paragraph *(c)* of this section for a labeling statement are effective on May 5, 1983 for each affected OTC drug product packaged for retail sale on or after that date, except that the requirement for a specific label reference to any identifying characteristics effective on February 6, 1984 for each affected OTC drug product packaged for retail sale on or after that date.

(3)   *Retail level effective date.* The tamper-resistant packaging requirement of paragraph *(b)* of this section is effective on February 6, 1984 for each affected OTC drug product held for sale on or after that date that was packaged for retail sale before May 5, 1983. This does not include the requirement in paragraph *(b)* of this section that the indicator or barrier to entry be distinctive by design. Products packaged for retail sale after May 5, 1983, are required to be in compliance with all aspects of the regulations without regard to the retail level effective date.

(Secs. 201(n), 501, 502, 505, 506, 507, 601, 602, 701, 52 Stat. 1049-1056 as amended, 55 Stat. 851, 59 Stat. 463 as amended (21 U.S.C. 321(n), 351, 352, 355, 356, 357, 361, 362, 371))

[47 FR 50449, Nov. 5, 1982; 48 FR 1707, Jan. 14, 1983, as amended at 48 FR 16664, Apr. 19, 1983; 48 FR 37624, Aug. 19, 1983; 48 FR 41579, Sept. 16, 1983]

Effective date note: paragraph *(g)*(3) of §211.132 was added at 47 FR 50449, Nov. 5, 1982, effective February 6, 1984. At 48 FR 41579, Sept. 16, 1983, FDA published an interim stay of the effective date of paragraph *(g)*(3).

§211.134   *Drug product inspection.*

(a)   Packaged and labeled products shall be examined during finishing operations to provide assurance that containers and packages in the lot have the correct label.

(b)   A representative sample of units shall be collected at the completion of finishing operations and shall be visually examined for correct labeling.

(c)   Results of these examinations shall be recorded in the batch production or control records.

§211.137   *Expiration dating.*

(a)   To assure that a drug product meets applicable standards of identity, strength, quality, and purity at the time of use, it shall bear an expiration date determined by appropriate stability testing described in §211.166.

(b)   Expiration dates shall be related to any storage conditions stated on the labeling, as determined by stability studies described in §211.166.

(c)   If the drug product is to be reconstituted at the time of dispensing, its labeling shall bear expiration information for both the reconstituted and unreconstituted drug products.

(d)   Expiration dates shall appear on labeling in accordance with the requirements of §201.17 of this chapter.

(e)   Homeopathic drug products shall be exempt from the requirements of this section.

(f)   Allergenic extracts that are labeled "No U.S. Standard of Potency" are exempt from the requirements of this section.

(g) Pending consideration of a proposed exemption, published in the FEDERAL REGISTER of September 29, 1978, the requirements in this section shall not be enforced for human OTC drug products if their labeling does not bear dosage limitations and they are stable for at least 3 years as supported by appropriate stability data.

(Secs. 502, 505, 512, 701, 52 Stat. 1050-1053 as amended, 1055-1056 as amended, 82 Stat. 343-349 (21 U.S.C. 352, 355, 360b, 371))

[43 FR 45077, Sept. 29, 1978, as amended at 46 FR 56412, Nov. 17, 1981]

Subpart H.   Holding and Distribution

§211.142   *Warehousing procedures.*

Written procedures describing the warehousing of drug products shall be established and followed. They shall include:

(a) Quarantine of drug products before release by the quality control unit.

(b) Storage of drug products under appropriate conditions of temperature, humidity, and light so that the identity, strength, quality, and purity of the drug products are not affected.

§211.150   *Distribution procedures.*

Written procedures shall be established, and followed, describing the distribution of drug products. They shall include:

(a) A procedure whereby the oldest approved stock of a drug product is distributed first. Deviation from this requirement is permitted if such deviation is temporary and appropriate.

(b) A system by which the distribution of each lot of drug product can be readily determined to facilitate its recall if necessary.

Subpart I.   Laboratory Controls

§211.160   *General requirements.*

(a) The establishment of any specifications, standards, sampling plans, test procedures, or other laboratory control mechanisms required by this subpart, including any change in such specifications, standards, sampling plans, test procedures, or other laboratory control mechanisms, shall be drafted by the appropriate organizational unit and reviewed and approved by the quality control unit. The requirements in this subpart shall be followed and shall be documented at the time of performance. Any deviation from the written specifications, standards, sampling plans, test procedures, or other laboratory control mechanisms shall be recorded and justified.

(b) Laboratory controls shall include the establishment of scientifically sound and appropriate specifications, standards, sampling plans, and test procedures designed to assure that components, drug product containers, closures, in-process materials, labeling, and drug products conform to appropriate standards of identity, strength, quality, and purity. Laboratory controls shall include:

(1) Determination of conformance to appropriate written specifications for the acceptance of each lot within each shipment of components, drug product containers, closures, and labeling used in the manufacture, processing, packing, or holding of drug products. The specifications shall include a description of the sampling and testing procedures used. Samples shall be representative and adequately identified. Such procedures shall also require appropriate retesting of any component, drug product container, or closure that is subject to deterioration.

(2)   Determination of conformance to written specifications and a description of sampling and testing procedures for in-process materials. Such samples shall be representative and properly identified.

(3)   Determination of conformance to written descriptions of sampling procedures and appropriate specifications for drug products. Such samples shall be representative and properly identified.

(4)   The calibration of instruments, apparatus, gauges, and recording devices at suitable intervals in accordance with an established written program containing specific directions, schedules, limits for accuracy and precision, and provisions for remedial action in the event accuracy and/or precision limits are not met. Instruments, apparatus, gauges, and recording devices not meeting established specifications shall not be used.

§211.165   *Testing and release for distribution.*

(a)   For each batch of drug product, there shall be appropriate laboratory determination of satisfactory conformance to final specifications for the drug product, including the identity and strength of each active ingredient, prior to release. Where sterility and/or pyrogen testing are conducted on specific batches of shortlived radio-pharmaceuticals, such batches may be released prior to completion of sterility and/or pyrogen testing, provided such testing is completed as soon as possible.

(b)   There shall be appropriate laboratory testing, as necessary, of each batch of drug product required to be free of objectionable microorganisms.

(c)   Any sampling and testing plans shall be described in written procedures that shall include the method of sampling and the number of units per batch to be tested; such written procedure shall be followed.

(d)   Acceptance criteria for the sampling and testing conducted by the quality control unit shall be adequate to assure that batches of drug products meet each appropriate specification and appropriate statistical quality control criteria as a condition for their approval and release. The statistical quality control criteria shall include appropriate acceptance levels and/or appropriate rejection levels.

(e)   The accuracy, sensitivity, specificity, and reproducibility of test methods employed by the firm shall be established and documented. Such validation and documentation may be accomplished in accordance with §211.194*(a)*(2).

(f)   Drug products failing to meet established standards or specifications and any other relevant quality control criteria shall be rejected. Reprocessing may be performed. Prior to acceptance and use, reprocessed material must meet appropriate standards, specifications, and any other relevant criteria.

§211.166   *Stability testing.*

(a)   There shall be a written testing program designed to assess the stability characteristics of drug products. The results of such stability testing shall be used in determining appropriate storage conditions and expiration dates. The written program shall be followed and shall include:

(1)   Sample size and test intervals based on statistical criteria for each attribute examined to assure valid estimates of stability;

(2)   Storage conditions for samples retained for testing;

(3)   Reliable, meaningful, and specific test methods;

(4)   Testing of the drug product in the same container-closure system as that in which the drug product is marketed;

124

(5) Testing of drug products for reconstitution at the time of dispensing (as directed in the labeling) as well as after they are reconstituted.

(b) An adequate number of batches of each drug product shall be tested to determine an appropriate expiration date and a record of such data shall be maintained. Accelerated studies, combined with basic stability information on the components, drug products, and container-closure system, may be used to support tentative expiration dates provided full shelf life studies are not available and are being conducted. Where data from accelerated studies are used to project a tentative expiration date that is beyond a date supported by actual shelf life studies, there must be stability studies conducted, including drug product testing at appropriate intervals, until the tentative expiration date is verified or the appropriate expiration date determined.

(c) For homeopathic drug products, the requirements of this section are as follows:

(1) There shall be a written assessment of stability based at least on testing or examination of the drug product for compatibility of the ingredients, and based on marketing experience with the drug product to indicate that there is no degradation of the product for the normal or expected period of use.

(2) Evaluation of stability shall be based on the same container-closure system in which the drug product is being marketed.

(d) Allergenic extracts that are labeled "No U.S. Standard of Potency" are exempt from the requirements of this section.

Secs. 502, 505, 512, 701, 52 Stat. 1050-1053 as amended, 1055-1056 as amended, 82 Stat. 343-349 (21 U.S.C. 352, 355, 360b, 371))

[43 FR 45077, Sept. 29, 1978, as amended at 46 FR 56412, Nov. 17, 1981]

§211.167 *Special testing requirements.*

(a) For each batch of drug product purporting to be sterile and/or pyrogen-free, there shall be appropriate laboratory testing to determine conformance to such requirements. The test procedures shall be in writing and shall be followed.

(b) For each batch of ophthalmic ointment, there shall be appropriate testing to determine conformance to specifications regarding the presence of foreign particles and harsh or abrasive substances. The test procedures shall be in writing and shall be followed.

(c) For each batch of controlled-release dosage form, there shall be appropriate laboratory testing to determine conformance to the specifications for the rate of release of each active ingredient. The test procedures shall be in writing and shall be followed.

§211.170 *Reserve samples.*

(a) An appropriately identified reserve sample that is representative of each lot in each shipment of each active ingredient shall be retained. The reserve sample consists of at least twice the quantity necessary for all tests required to determine whether the active ingredient meets its established specifications, except for sterility and pyrogen testing. The retention time is as follows:

(1) For an active ingredient in a drug product other than those described in paragraph *(a)* (2) and (3) of this section, the reserve sample shall be retained for one year after the expiration date of the last lot of the drug product containing the active ingredient.

(2)   For an active ingredient in a radioactive drug product, except for nonradioactive reagent kits, the reserve sample shall be retained for:

   (i)   Three months after the expiration date of the last lot of the drug product containing the active ingredient if the expiration dating period of the drug product is 30 days or less; or

   (ii)   Six months after the expiration date of the last lot of the drug product containing the active ingredient if the expiration dating period of the drug product is more than 30 days.

(3)   For an active ingredient in an OTC drug product that is exempt from bearing an expiration date under §211.137, the reserve sample shall be retained for three years after distribution of the last lot of the drug product containing the active ingredient.

(b)   An appropriately identified reserve sample that is representative of each lot or batch of drug product shall be retained and stored under conditions consistent with product labeling. The reserve sample shall be stored in the same immediate container-closure system in which the drug product is marketed or in one that has essentially the same characteristics. The reserve sample consists of at least twice the quantity necessary to perform all the required tests, except those for sterility and pyrogens. Reserve samples, except those drugs products described in paragraph *(b)* (2), shall be examined visually at least once a year for evidence of deterioration unless visual examination would affect the integrity of the reserve samples. Any evidence of reserve sample deterioration shall be investigated in accordance with §211.192. The results of the examination shall be recorded and maintained with other stability data on the drug product. Reserve samples of compressed medical gases need not be retained. The retention time is as follows:

(1)   For a drug product other than those described in paragraphs *(b)* (2) and (3) of this section, the reserve sample shall be retained for one year after the expiration date of the drug product.

(2)   For a radioactive drug product, except for nonradioactive reagent kits, the reserve sample shall be retained for:

   (i)   Three months after the expiration date of the drug product if the expiration dating period of the drug product is 30 days or less; or

   (ii)   Six months after the expiration date of the drug product if the expiration dating period of the drug product is more than 30 days.

(3)   For an OTC drug product that is exempt for bearing an expiration date under §211.137, the reserve sample must be retained for three years after the lot or batch of drug product is distributed.

(Secs. 501, 502, 505, 512, 701, 52 Stat. 1049-1053 as amended, 1055-1056 as amended, 82 Stat. 343-351 (21 U.S.C. 351, 352, 355, 360b, 371))

[48 FR 13025, Mar. 29, 1983]

§211.173   *Laboratory animals.*

Animals used in testing components, in-process materials, or drug products for compliance with established specifications shall be maintained and controlled in a manner that assures their suitability for their intended use. They shall be identified, and adequate records shall be maintained showing the history of their use.

§211.176   *Penicillin contamination.*

If a reasonable possibility exists that a non-penicillin drug product has been exposed to cross-contamination with penicillin, the non-penicillin drug product shall be tested for the presence of penicillin. Such drug product shall not be marketed if detectable levels are found when tested according to procedures specified in 'Procedures for Detecting and Measuring Penicillin Contamination in Drugs,' which is corporated by reference. Copies are available from the Division of Drug Biology (HFN-170), Center for Drugs and Biologics, Food and Drug Administration, 200 C St. S.W., Washington, D.C. 20204, or available for inspection at the Office of the Federal Register, 1100 L St. N.W., Washington, D.C. 20408.

[43 FR 45077, Sept. 29, 1978, as amended at 47 FR 9396, Mar. 5, 1982; 50 FR 8996, Mar. 8, 1985]

Subpart J.   Records and Reports

§211.180   *General requirements.*

(a)   Any production, control, or distribution record that is required to be maintained in compliance with this part and is specifically associated with a batch of a drug product shall be retained for at least one year after the expiration date of the batch or, in the case of certain OTC drug products lacking expiration dating because they meet the criteria for exemption under §211.137, three years after distribution of the batch.

(b)   Records shall be maintained for all components, drug product containers, closures, and labeling for at least one year after the expiration date or, in the case of certain OTC drug products lacking expiration dating because they meet the criteria for exemption under §211.137, three years after distribution of the last lot of drug product incorporating the component or using the container, closure, or labeling.

(c)   All records required under this part, or copies of such records, shall be readily available for authorized inspection during the retention period at the establishment where the activities described in such records occurred. These records or copies thereof shall be subject to photocopying or other means of reproduction as part of such inspection. Records that can be immediately retrieved from another location by computer or other electronic means shall be considered as meeting the requirements of this paragraph.

(d)   Records required under this part may be retained either as original records or as true copies such as photocopies, microfilm, microfiche, or other accurate reproductions of the original records. Where reduction techniques, such as microfilming, are used, suitable reader and photocopying equipment shall be readily available.

(e)   Written records required by this part shall be maintained so that data therein can be used for evaluating, at least annually, the quality standards of each drug product to determine the need for changes in drug product specifications or manufacturing or control procedures. Written procedures shall be established and followed for such evaluations and shall include provisions for:

(1)   A review of every batch, whether approved or rejected, and, where applicable, records associated with the batch.

(2)   A review of complaints, recalls, returned or salvaged drug products, and investigations conducted under §211.192 for each drug product.

(f)   Procedures shall be established to assure that the responsible officials of the firm, if they are not personally involved in or immediately aware of such actions, are notified in writing of any investigations conducted under §§211.198, 211.204, or 211.208 of these regulations, any recalls, reports of inspectional observations issued by the Food and Drug Administration, or any regulatory actions relating to good manufacturing practices brought by the Food and Drug Administration.

§211.182 *Equipment cleaning and use log.*

A written record of major equipment cleaning, maintenance (except routine maintenance such as lubrication and adjustments), and use shall be included in individual equipment logs that show the date, time, product, and lot number of each batch processed. If equipment is dedicated to manufacture of one product, then individual equipment logs are not required, provided that lots or batches of such product follow in numerical order and are manufactured in numerical sequence. In cases where dedicated equipment is employed, the records of cleaning, maintenance, and use shall be part of the batch record. The persons performing and double-checking the cleaning and maintenance shall date and sign or initial the log indicating that the work was performed. Entries in the log shall be in chronological order.

§211.184 *Component, drug product container, closure, and labeling records.*

These records shall include the following:

(a) The identity and quantity of each shipment of each lot of components, drug product containers, closures, and labeling; the name of the supplier; the supplier's lot number(s) if known; the receiving code as specified in §211.80; and the date of receipt. The name and location of the prime manufacturer, if different from the supplier, shall be listed if known.

(b) The results of any test or examination performed (including those performed as required by §211.82*(a)*, §211.84*(d)*, or §211.122*(a)*) and the conclusions derived therefrom.

(c) An individual inventory record of each component, drug product container, and closure and, for each component, a reconciliation of the use of each lot of such component. The inventory record shall contain sufficient information to allow determination of any batch or lot of drug product associated with the use of each component, drug product container, and closure.

(d) Documentation of the examination and review of labels and labeling for conformity with established specifications in accord with §211.122*(c)* and 211.130*(c)*.

(e) The disposition of rejected components, drug product containers, closure, and labeling.

§211.186 *Master production and control records.*

(a) To assure uniformity from batch to batch, master production and control records for each drug product, including each batch size thereof, shall be prepared, dated, and signed (full signature, handwritten) by one person and independently checked, dated, and signed by a second person. The preparation of master production and control records shall be described in a written procedure and such written procedure shall be followed.

(b) Master production and control records shall include:

(1) The name and strength of the product and a description of the dosage form;

(2) The name and weight or measure of each active ingredient per dosage unit or per unit of weight or measure of the drug product, and a statement of the total weight or measure of any dosage unit;

(3) A complete list of components designated by names or codes sufficiently specific to indicate any special quality characteristic;

(4) An accurate statement of the weight or measure of each component, using the same weight system (metric, avoirdupois, or apothecary) for each component. Reasonable variations may be permitted, however, in the amount of components necessary for the preparation in the dosage form, provided they are justified in the master production and control records;

(5) A statement concerning any calculated excess of component;

(6) A statement of theoretical weight or measure at appropriate phases of processing;

(7) A statement of theoretical yield, including the maximum and minimum percentages of theoretical yield beyond which investigation according to §211.192 is required;

(8) A description of the drug product containers, closures, and packaging materials, including a specimen or copy of each label and all other labeling signed and dated by the person or persons responsible for approval of such labeling;

(9) Complete manufacturing and control instructions, sampling and testing procedures, specifications, special notations, and precautions to be followed.

§211.188 *Batch production and control records.*

Batch production and control records shall be prepared for each batch of drug product produced and shall include complete information relating to the production and control of each batch. These records shall include:

(a) An accurate reproduction of the appropriate master production or control record, checked for accuracy, dated, and signed;

(b) Documentation that each significant step in the manufacture, processing, packing, or holding of the batch was accomplished, including:

(1) Dates;

(2) Identity of individual major equipment and lines used;

(3) Specific identification of each batch of component or in-process material used;

(4) Weights and measures of components used in the course of processing;

(5) In-process and laboratory control results;

(6) Inspection of the packaging and labeling area before and after use;

(7) A statement of the actual yield and a statement of the percentage of theoretical yield at appropriate phases of processing;

(8) Complete labeling control records, including specimens or copies of all labeling used;

(9) Description of drug product containers and closures;

(10) Any sampling performed;

(11) Identification of the persons performing and directly supervising or checking each significant step in the operation;

(12) Any investigation made according to §211.192;

(13) Results of examinations made in accordance with §211.134.

§211.192 *Production record review.*

All drug product production and control records, including those for packaging and labeling, shall be reviewed and approved by the quality control unit to determine compliance with all established, approved written procedures before a batch is released or distributed. Any unexplained discrepancy (including a percentage of theoretical yield exceeding the maximum or minimum percentages established in master production and control records) or the failure of a batch or any of its components to meet any of its specifications shall be thoroughly investigated, whether or not the batch has already been distributed. The investigation shall extend to other batches of the same drug product and other drug products that may have been associated with the specific failure or discrepancy. A written record of the investigation shall be made and shall include the conclusions and follow-up.

§211.194   *Laboratory records.*

(a)   Laboratory records shall include complete date derived from all tests necessary to assure compliance with established specifications and standards, including examinations and assays, as follows:

(1)   A description of the sample received for testing with identification of source (that is, location from where sample was obtained), quantity, lot number or other distinctive code, date sample was taken, and date sample was received for testing.

(2)   A statement of each method used in the testing of the sample. The statement shall indicate the location of data that establish that the methods used in the testing of the sample meet proper standards of accuracy and reliability as applied to the product tested. (If the method employed is in the current revision of the United States Pharmacopeia, National Formulary, Association of Official Analytical Chemists, Book of Methods,[2] or in other recognized standard referenced, or is detailed in an approved new drug application and the reference method is not modified, a statement indicating the method and reference will suffice.) The suitability of all testing methods used shall be verified under actual conditions of use.

(3)   A statement of the weight or measure of sample used for each test, where appropriate.

(4)   A complete record of all data secured in the course of each test, including all graphs, charts, and spectra from laboratory instrumentation, properly identified to show the specific component, drug product container, closure, in-process material, or drug product, and lot tested.

(5)   A record of all calculations performed in connection with the test, including units of measure, conversion factors, and equivalency factors.

(6)   A statement of the results of tests and how the results compare with established standards of identity, strength, quality, and purity for the component, drug product container, closure, in-process material, or drug product tested.

(7)   The initials or signature of the person who performs each test and the date(s) the tests were performed.

(8)   The initials or signature of a second person showing that the original records have been reviewed for accuracy, completeness, and compliance with established standards.

(b)   Complete records shall be maintained of any modification of an established method employed in testing. Such records shall include the reason for the modification and data to verify that the modification produced results that are at least as accurate and reliable for the material being tested as the established method.

(c)   Complete records shall be maintained of any testing and standardization of laboratory reference standards, reagents, and standard solutions.

(d)   Complete records shall be maintained of the periodic calibration of laboratory instruments, apparatus, gauges, and recording devices required by §211.160*(b)*(4).

(e)   Complete records shall be maintained of all stability testing performed in accordance with §211.166.

---

[2]Copies may be obtained from: Association of Official Analytical Chemists, P.O. Box 540, Benjamin Franklin Station, Washington, D.C. 20204.

§211.196 *Distribution records.*

Distribution records shall contain the name and strength of the product and description of the dosage form, name and address of the consignee, date and quantity shipped, and lot or control number of the drug product. For compressed medical gas products, distribution records are not required to contain lot or control numbers.

(Approved by the Office of Management and Budget under control number 0910-0139)

(Secs. 501, 502, 512, 701, 52 Stat. 1049-1051 as amended, 1055-1056 as amended, 82 Stat. 343-351 (21 U.S.C. 351, 352, 360b, 371))

[49 FR 9865, Mar. 16, 1984]

§211.198 *Complaint files.*

(a) Written procedures describing the handling of all written and oral complaints regarding a drug product shall be established and followed. Such procedures shall include provisions for review by the quality control unit, of any complaint involving the possible failure of a drug product to meet any of its specifications and, for such drug products, a determination as to the need for an investigation in accordance with §211.192.

(b) A written record of each complaint shall be maintained in a file designated for drug product complaints. The file regarding such drug product complaints shall be maintained at the establishment where the drug product involved was manufactured, processed, or packed, or such file may be maintained at another facility if the written records in such files are readily available for inspection at that other facility. Written records involving a drug product shall be maintained until at least one year after the expiration date of the drug product, or one year after the data that the complaint was received, whichever is longer. In the case of certain OTC drug products lacking expiration dating because they meet the criteria for exemption under §211.137, such written records shall be maintained for three years after distribution of the drug product.

(1) The writen record shall include the following information, where known: the name and strength of the drug product, lot number, name of complainant, nature of complaint, and reply to complainant.

(2) Where an investigation under §211.192 is conducted, the written record shall include the findings of the investigation and follow-up. The record or copy of the record of the investigation shall be maintained at the establishment where the investigation occurred in accordance with §211.180(c).

(3) Where an investigation under §211.192 is not conducted, the written record shall include the reason that an investigation was found not to be necessary and the name of the responsible person making such a determination.

Subpart K. Returned and Salvaged Drug Products

§211.204 *Returned drug products.*

Returned drug products shall be identified as such and held. If the conditions under which returned drug products have been held, stored, or shipped before or during their return, of if the condition of the drug product, its container, carton, or labeling, as a result of storage or shipping, casts doubt on the safety, identity, strength, quality or purity of the drug product, the returned drug product shall be destroyed unless examination, testing, or other investigations prove the drug product meets appropriate standards of safety, identity, strength, quality, or purity. A drug product may be reprocessed provided the subsequent drug product meets appropriate standards, specifications, and characteristics. Records of returned drug products shall be maintained and shall include the name and label potency of the drug product dosage form, lot number (or control number or batch number), reason for the return, quantity returned, date of disposition, and ultimate disposition of the returned drug product. If the reason for a drug product being returned implicates associated batches, an appropriate investigation shall be conducted in accordance with the requirements of §211.192. Procedures for the holding, testing, and reprocessing of returned drug products shall be in writing and shall be followed.

§211.208   *Drug product salvaging.*

Drug products that have been subjected to improper storage conditions including extremes in temperature, humidity, smoke, fumes, pressure, age, or radiation due to natural disasters, fires, accidents, or equipment failures shall not be salvaged and returned to the marketplace. Whenever there is a question whether drug products have been subjected to such conditions, salvaging operations may be conducted only if there is *(a)* evidence from laboratory tests and assays (including animal feeding studies where applicable) that the drug products meet all applicable standards of identity, strength, quality, and purity and *(b)* evidence from inspection of the premises that the drug products and their associated packaging were not subjected to improper storage conditions as a result of the disaster or accident. Organoleptic examinations shall be acceptable only as supplemental evidence that the drug products meet appropriate standards of identity, strength, quality, and purity. Records including name, lot number, and disposition shall be maintained for drug products subject to this section.

Annex III

PLANT WORKING GROUP GUIDELINES FOR PROPOSAL SUBMISSION

These annotated guidelines are presented for consideration by prospective proposal submitters to facilitate the process of approval. The PWG has found that the proposals so far submitted for their consideration have omitted information that is considered minimal and essential for their approval. Basically, the group would like to see detailed objectives, materials and methods, including methodology for monitoring the experiments, and expected results. At least summary data should be submitted to support the proposal. A check list of detailed requirements should include, but is not limited to:

1. Give common and scientific names of plants and cultivars, if appropriate. "Tomato plants will be inoculated" is insufficient.

2. *If appropriate,* give data or information on the relative homogeneity of the plant cultivar, *and specific genetic markets the cultivar is known to possess.*

3. Give specific strain designations of those you expect to use. "Some strains of *Agrobacterium rhizogenes* will be used . . ." is insufficient.

4. Give the method(s) by which the vector will be or is constructed. Diagrams are very helpful and may be necessary for adequate understanding of the construct. Explain the advantages (and disadvantage(s), if appropriate) of your vectors, if other candidate vectors could be considered.

5. If live host microorganisms are used to introduce vectors or are vectors themselves, indicate how they compare with wild-type strains. If disabled pathogens *or vectors* are used as hosts, indicate measures that will most likely prevent these microorganisms from regaining *or acquiring* pathogenic potential.

6. Give criteria and methods by which the host microorganism will be monitored. If live host microorganisms are required to be present in field trials, indicate the means of strain identification and retrieval. *If microorganisms are used to introduce vectors, the assessment of subsequent absence of the microorganisms should be specified.*

7. If the microorganisms are transfer deficient, provide some documentation either via the proposal or appropriate references.

8. If the vectors are transfer deficient, provide some documentation via the proposal or appropriate references.

9. If the vector is likely to survive independently of the hosts, refer to this possibility; if the answer is in the realm of reasonably high probability, provide data to assess such transfer to likely microorganisms.

10. Provide data from greenhouse and/or growth chamber studies under simulated field conditions to support prospective field studies.

11. If appropriate, provide data on engineered plants and controls fed to test animals, such as mice.

12. Provide data for field plot design on the following:

    a. Total area;

    b. Location: where, how many;

    c. Plot design: replication, row spacing, planting, border rows etc.;

    d. Name cultivar(s), if appropriate;

    e. Specify plant monitoring procedures: frequency; types of data to be *obtained,* including leaf, seed, fruit, *or,* root *characteristics;* abnormalities, such as diseases; insect population monitoring; collection of meteorological data etc.; types of data to be sought, such as yield, resistance to *stress,* lodging etc.;

    f. Specify monitoring of the vector and/or introduced DNA;

    g. Specify access and security measures.

13 List qualifications of people involved in the experiments.

Note: Appendix L (Release into the environment of certain plants, Federal Register 48: 24549 and 48: 37199) should be consulted for general guidelines regarding plants.

Annex IV

POINTS TO CONSIDER FOR ENVIRONMENTAL TESTING OF MICROORGANISMS

DEPARTMENT OF HEALTH AND HUMAN SERVICES

National Institutes of Health

RECOMBINANT DNA ADVISORY COMMITTEE; MEETING

Pursuant to Pub. L 92-463, notice is hereby given of a meeting of the Recombinant DNA Advisory Committee at the National Institutes of Health. Building 31C. Conference Room 6, 9000 Rockville Pike, Bethesda, Maryland 20205, on May 3, 1985 from 9:00 a.m. to adjournment at approximately 5:00 p.m. This meeting will be open to the public to discuss:

Report of the Working Group on Release into the environment;
Report of the Working Group on Human Gene Therapy;
Proposed co-ordinated framework for regulation of biotechnology weapons;
Proposed working group on biological weapons;
Amendment of Guidelines; and
Other matters to be considered by the Committee.

Attendance by the public will be limited to space available. Members of the public wishing to speak at the meeting may be given such opportunity at the discretion of the chair.

Dr. William J. Gartland, Jr., Executive Secretary, Recombinant DNA Advisory Committee, National Institutes of Health, Building 31, Room 3B10, telephone (301) 496-6051, will provide materials to be discussed at the meeting, rosters of committee members, substantive program information. A summary of the meeting will be available at a later date.

Dated: 8 March 1985.

BETTY J. BEVERIDGE.

Committee Management Officer, NIH

OMB's "Mandatory Information Requirements for Federal Assistance Program Announcements" (45 FR 39592) requires a statement concerning the official government programs contained in the *Catalog of Federal Domestic Assistance.* Normally NIH lists in its announcements the number and title of affected individual programs for the guidance of the public. Because the guidance in this notice covers not only virtually every NIH program but also essentially every federal research program in which DNA recombinant molecule techniques could be used. It has been determined to be not cost effective or in the public interest to attempt to list these programs. Such a list would likely require several additional pages. In addition, NIH could not be certain that every federal program would be included as many federal agencies, as well as private organizations, both national and international, have elected to follow the NIH

Guildelines. In lieu of the individual program listing, NIH invites readers to direct questions to the information address above about whether individual programs listed in the *Catalog of Federal Domestic Assistance* are affected.

(FR Doc. 85-7063 Filed 3-27-85: 8:45 a.m.)

RECOMBINANT DNA RESEARCH;
PROPOSED ACTIONS UNDER GUIDELINES

Agency: National Institutes of Health. PHS, HHS

Action: Notice of Proposed Actions under NIH Guidelines for Research Involving Recombinant DNA Molecules

Summary: This notice sets forth proposed actions to be taken under the NIH Guidelines for Research Involving Recombinant DNA Molecules. Interested parties are invited to submit comments concerning these proposals. After consideration of these proposals and comments by the NIH Recombinant DNA Advisory Committee (RAC) at its meeting on May 3, 1985, the Director of the National Institutes of Health will issue decisions on these proposals in accord with the Guidelines.

Date: Comments must be received by April 29, 1985.

Address: Written comments and recommendations should be submitted to the Director, Office of Recombinant DNA Activities, Building 31, Room 3B10, National Institutes of Health, Bethesda, Maryland 20205. All comments received in timely response to this notice will be considered and will be available for public inspection in the above office on weekdays between the hours of 8:30 a.m. and 5:00 p.m. Comments received by close of business April 26, 1985, will be reproduced and distributed to the RAC for consideration at its May 3, 1985, meeting.

For further information contact:

Background documentation and additional information can be obtained from Drs. Stanley Barban and Elizabeth Milewski, Office of Recombinant DNA Activities, National Institutes of Health, Bethesda, Maryland 20205 (301) 496-6051.

Supplementary information: The National Institutes of Health will consider the following actions under the Guidelines for Research Involving Recombinant DNA Molecules.

I.   PROPOSED POINTS TO CONSIDER FOR ENVIRONMENTAL TESTING OF MICROORGANISMS

Deliberate release into the environment of any organism containing recombinant DNA, except certain plants as described in Appendix L, falls under Section III-A of the NIH Guidelines. Experiments in this category cannot be initiated without submission of relevant information on the proposed experiments to NIH, review by the RAC after publication for public comment, and specific approval by NIH.

The RAC Working Group on Release into the Environment has prepared draft submission guidelines for individuals preparing proposals involving testing in the environment of microorganisms derived by recombinant DNA techniques. The proposed guidance follows:

Points to Consider for Submissions involving Testing in the Environment of Microorganisms Derived by Recombinant DNA Techniques

Experiments in this category require specific review by the Recombinant DNA Advisory Committee (RAC) and approvals by the National Institutes of Health (NIH) and the Institutional Biosafety Committee (IBC) before initiation. The IBC is expected to make an independent evaluation although this evaluation need not occur before consideration of an experiment by the RAC. Relevant information on the proposed experiments should be submitted to the Office of Recombinant DNA Activities (ORDA). The objective of this review procedure is to evaluate the potential environmental effects of testing of microorganisms that have been modified by recombinant DNA techniques.

These following points to consider have been developed by the RAC Working Group on Release into the Environment as a suggested list for scientists preparing proposals on environmental testing of microorganisms, including viruses, that have been modified using recombinant DNA techniques. The review of proposals for environmental testing of modified organisms is being done on a case-by-case basis because the range of possible organisms, applications, and environments indicate that no standard set of procedures is likely to be appropriate in all circumstances. However, some common considerations allow the construction of points to consider such as those below. *Information on all these points will not be necessary in all cases but will depend on the properties of the parental organism and the effect of the modification on these properties*

Approval of small-scale field tests will depend upon the results of laboratory and greenhouse testing of the properties of the modified organism. We anticipate that monitoring of small-scale field tests will provide data on environmental effects of the modified organism. Such data may be a necessary part of the consideration of requests for approval of large-scale tests and commercial applications.

I.   Summary

Present a summary of the proposed trial including objectives, significance, and justification for the request.

II.   Genetic Considerations of Modified Organism to be Tested

A.   *Characteristics of the nonmodified Parental Organism*

1.   Information on identification, taxonomy, source, and strain.

2.   Information on organism's reproductive cycle and capacity for genetic transfer.

B.   *Molecular Biology of the Modified Organism*

1.   Introduced Genes

  a.   Source and function of the DNA sequence used to modify the organism to be tested in the environment.

  b.   Identification, taxonomy, source and strain or organism donating the DNA.

2.   Construction of the Modified Organism

  a.   Describe the method(s) by which the vector with insert(s) has been constructed. Include diagrams as appropriate.

  b.   Describe the method of introduction of the vector carrying the insert into the organism to be modified and the procedure for selection of the modified organism.

c. Specify the amount and nature of any vector and/or donor DNA remaining in the modified organism.

d. Give the laboratory containment conditions specified by the NIH Guidelines for the modified organism.

3. Genetic Stability and Expression

Present results and interpretation of preliminary tests designed to measure genetic stability and expression of the introduced DNA in the modified organism.

III Environmental Considerations

The intent of gathering ecological information is to assess to the effects of survival, reproduction, and/or dispersal of the modified organism. For this purpose, information should be provided where possible and appropriate on: (i) Relevant ecological characteristics of the nonmodified organism; (ii) the corresponding characteristics of the modified organism; and (iii) the physiological and ecological role of donated genetic sequences in the donor and in the modified organism(s). For the following points, provide information where possible and appropriate on the nonmodified organism and a prediction of any change that may be elicited by the modification.

A. Habitat and Geographic Distribution

B. Physical and Chemical factors which can affect Survival, Reproduction, and Dispersal

C. Biological Interactions

1. Host range.

2. Interactions with and effects on other organisms in the environment including effects on competitors, prey, hosts, symbionts, predators, parasites, and pathogens.

3. Pathogenicity, infectivity, toxicity, virulence, or as a carrier (vector) of pathogens.

4. Involvement in biogeochemical or in biological cycling processes (e.g. mineral cycling, cellulose and lignin degradation, nitrogen fixation, pesticide degradation).

5. Frequency with which populations undergo shifts in important ecological characteristics such as those listed in III-C points 1 through 4 above.

6. Likelihood of exchange of genetic information between the modified organism and other organisms in nature.

IV. Proposed Field Trials

A. Pre-Field Trial Considerations

Provide data related to any anticipated effects of the modified microorganism on target and nontarget organisms from microcosm, greenhouse, and/or growth chamber experiments that simulate trial conditions. The methods of detection and sensitivity of sampling techniques and periodicity of sampling should be indicated. These studies should include, where relevant, assessment of the following items:

1. Survival of the modified organism.

2. Replication of the modified organism.

3. Dissemination of the modified organism by wind, water, soil, mobile organisms, and other means.

138

B. Conditions of the Trial

Describe the trial involving release of the modified organism into the environment:

1. Numbers of organisms and methods of application.

2. Provide information including diagrams of the experimental location and the immediate surroundings. Describe characteristics of the site that would influence containment or dispersal.

3. If the modified organism has a target organism, provide the following:

    a. Identification and taxonomy.

    b. The anticipated mechanism and result of the interaction between the released microorganism and the target organism.

C. Containment

Indicate containment procedures in the event of accidental release as well as intentional release and procedures for emergency termination of the experiment. Specify access and security measures for the area(s) in which the tests will be performed.

D. Monitoring

Describe monitoring procedures and their limits of detection for survival, dissemination, and nontarget interactions of the modified microorganism. Include periodicity of sampling and rationale for monitoring procedures. Collect data to compare the modified organisms with the nonmodified microorganism most similar to the modified organism at the site of the trial. Results of monitoring should be submitted to the RAC according to a schedule specified at the time of approval.

V. Risk Analysis

Results of testing in artificial contained environments together with careful consideration of the genetics, biology, and ecology of the nonmodified and the modified organisms will enable a reasonable prediction of whether or not significant risk of environmental damage will result from the release of the modified organism in the small-scale field test. In this section, the information requested in Sections II, III, and IV should be summarized to present an analysis of possible risks to the environment in the test as it is proposed. The issues addressed might include but not be limited to the following items:

A. The Nature of the Organism

1. The role of the nonmodified organism in the environment of the test site, including any adverse effects on other organisms.

2. Evaluation of whether or not the specific genetic modification (e.g. deletion, insertion, modification of specific DNA sequences) would alter the potential for significant adverse effects.

3. Evaluation of results of tests conducted in contained environments to predict the ecological behavior of the modified organism relative to that of its nonmodified parent.

B. The Nature of the Test

Discuss the following specific features of the experiment that are designed to minimize potential adverse effects of the modified organism:

1. Test site location and area.

2. Introduction protocols.

3. Numbers of organisms and their expected reproductive capacity.

4. Emergency procedures for aborting the experiment.

5. Procedures conducted at the termination of the experiment.

II. PROPOSED REVISION OF APPENDIX C

Dr. Jack J. Manis of the Upjohn Company, Kalamazoo, Michigan, has proposed that the following kinds of experiments be made exempt under Section III-D-5 and the following language be included in Appendix C of the NIH Guidelines.

Experiments and processes utilizing recombinant DNA containing derivatives of *Streptomyces fradioe* or *Streptomyces lincolnensis* are exempt from the Guidelines at all levels of volume scale when the recombinant DNA molecules contained in these hosts are derived solely from nonpathogenic streptomycetes. The nonpathogenicities of the recombinant DNA sources are determined by the local IBC.

For these exempt laboratory experiments, BL1 physical containment conditions are recommended.

For large-scale fermentation experiments BL1-LS physical containment conditions are recommended. However, following review by the IBC of appropriate data for a particular host-vector system some latitude in the application of BL1-LS requirements as outlined in Appendix K-II-A through K-II-F is permitted.

Exceptions. Experiments described in Section III-A which require specific RAC review and NIH approval before initiation of the experiment.

Large-scale experiments (e.g. more than 10 liters of culture) require prior IBC review and approval (see Section III-B-5).

Explanation of this proposed modification if provided in the submission.

III. PROPOSED AMENDMENT OF PART III

In a memorandum dated February 12, 1985, Dr. Bernard Talbot, Deputy Director of the National Institute of Allergy and Infectious Diseases, noted that under the NIH Guidelines for Research Involving Recombinant DNA Molecules certain proposals are received by NIH for review by the NIH Recombinant DNA Advisory Committee (RAC) and subsequent NIH approval. These include proposals which are required to be submitted from institutions that receive support for recombinant DNA research from NIH, and also proposals voluntarily submitted by institutions that receive no NIH support for recombinant DNA research. Recently other Federal agencies have made steps toward assuming new roles in review of recombinant DNA proposals.

Because of these developments, it could now happen that a proposal submitted to the NIH for RAC review and NIH approval (either from an institution that receives NIH funding for recombinant DNA research or voluntarily submitted by an institution that receives no such support) may be also submitted to another Federal agency for review.

Dr. Talbot states:

In such a case, I believe it would be advantageous for NIH to have the option of deferring to the review and approval by the other Federal agency rather than always going through review and approval by both the other Federal agency and the NIH. In order to allow this, I request that the following proposed change in the NIH Guidelines be issued for public comment, and placed on the agenda of the next RAC meeting. I propose that a new sentence be added at the end of Section III-A of the Guidelines ("Experiments that Require RAC Review and NIH and IBC Approval Before Initiation") just before Section III-A-1 of the Guidelines, as follows: "If experiments in this category are submitted for review to another Federal agency, the submitter shall notify ORDA: ORDA may then determine that such review serves the same purpose, and based on that determination, notify the submitter that no RAC review will take place, no NIH approval is necessary, and the experiment may proceed upon approval from the other Federal Agency."

Additional background information is provided in the memorandum.

IV. PROPOSED RAC WORKING GROUP

Messrs. Lee Rogers and Jeremy Rifkin of the Foundation on Economic Trends, Washington, D.C., submitted the following letter dated February 28, 1985, to NIH:

We are formally proposing that the Recombinant DNA Advisory Committee (RAC) of the National Institutes of Health (NIH) establish a working subgroup whose stated purpose would be to examine potential uses of recombinant DNA technology for offensive and defensive biological weapons systems. In addition, this subgroup will also explore current Department of Defense (DOD) programs specifically designed to develop "defensive" preparedness against the threat of genetic engineering warfare by aggressor nations or terrorists. It should be made clear that such a study is designed to look into the potential as well as actual uses of recombinant DNA technology for military purposes regardless of whether such experimentation is being conducted at this time. The working subgroup on biological warfare will make its findings available to the RAC, NIH, and interested members of the public. The working subgroup may also wish to make recommendations regarding future oversight of recombinant DNA work in this field.

It is no longer possible to ignore the potential military uses of recombinant DNA experimentation in light of the DOD's plan to construct an aerosol test laboratory at Dugway Proving Ground in Utah. The military has stated its intention to use this lab to test defenses against biological warfare experiments and it further stated that it will be working with deadly biological pathogens. In November 1984, the Secretary of Defense, Caspar Weinberger, stated in a letter to Senator Jim Sasser: "We continue to obtain new evidence that the Soviet Union has maintained its offensive biological warfare program and that it is exploring genetic engineering to expand their program's scope. Consequently, it is essential and urgent that we develop and field adequate biological and toxin protection." (See enclosed document.) In light of these recent developments, it is imperative that the RAC begin a thorough and comprehensive study of the potential uses of recombinant DNA technology for military purposes.

Since its inception, RAC has involved itself in every aspect of recombinant DNA technology in an effort to develop procedures, guidelines, protocols, and ethical standards to oversee this research. The only area of recombinant DNA experimentation that has not yet been rigorously examined is the potential military uses. Therefore, I would think that this committee would find it helpful to explore the potential military uses of recombinant DNA technology in order to facilitate a better understanding of the various issues involved. Moreover, it is altogether appropriate for the RAC to engage in such a study as the DOD has stated on many occasions that it is adhering to the guidelines established by this committee and the NIH. An independent study by the RAC of the military potential or recombinant DNA technology can only serve to better inform the Executive Branch, Congress and the public of the issues involved in this particular field.

Dated: 11 March 1965.

ANTHONY S. FAUCI.

Director,
National Institute of Allergy and Infectious Diseases,
National Institutes of Health

OMB's "Mandatory Information Requirements for Federal Assistance Program Announcements" (45 FR 39592) requires a statement concerning the official government programs contained in the *Catalog of Federal Domestic Assistance.* Normally NIH lists in its announcements the number and title of affected individual programs for the guidance of the public. Because the guidance in this notice covers not only virtually every NIH program but also essentially every federal research program in which DNA recombinant molecule techniques could be used, it has been determined to be not cost effective or in the public interest to attempt to list these programs. Such a list would likely require several additional pages. In addition. NIH could not be certain that every federal program would be included as many federal agencies, as well as private organizations, both national and international, have elected to follow the NIH Guidelines. In lieu of the individual program listing, NIH invites readers to direct questions to the information address above about whether individual programs listed in the Catalog of Federal Domestic Assistance are affected.

(FR Doc. 85-7064 Filed 3-27-85; 8:45 am)
BILLING CODE 4140-01-M

142

Glossary

Asporogenic	Without spores (nonspore forming)
Biotechnology	A collection of processes and techniques that involve the use of living organisms or substances from those organisms to make or modify products from raw materials for industrial, agricultural or medical purposes
B. subtilis *(Bacillus subtilis)*	An aerobic spore-forming cylindrical organism, a bacilli, seldom pathogenic except in immunologically compromised hosts
Cell fusion	Combining nuclei and cytoplasm from different cells to form a single hybrid cell
Clone	A group of genetically identical cells or organisms asexually descended from a common ancestor. All cells in the clone have the same genetic material and are exact copies of the original
Conjugation	The one-way transfer of DNA between bacteria in cellular contact
DNA	Deoxyribonucleic acid; the carrier of genetic information found in all living organisms. Every inherited characteristic is coded somewhere in an organism's complement of DNA
E. coli *(Escherichia coli)*	A species of bacteria that commonly inhabits the human intestine and the intestinal tract of most other vertebrates as well
Ecosystem	A term used to describe the total ecology of an environment
Eucaryote	A cell or organism with membrane-bound, structurally discrete nuclei and well-developed cell organelles. Eucaryotes include all organisms except viruses, bacteria and blue-green algae. (See procaryote)
Exotic organism	Non-indigenous; an organism which is placed into an environment in which it normally is not found
Expression	The translation of a gene's DNA sequence by RNA into protein
Fermentation	An anaerobic bioprocess in which yeasts, bacteria and molds are used to convert raw materials into products such as alcohol, acid and cheese
Gene splicing	The use of site-specific enzymes that cleave and reform bonds in DNA to create modified DNA sequences

143

Genetic engineering	A collection of technologies used to alter the hereditary apparatus of a living cell enabling the cell to produce more or different chemicals or to perform completely different functions. These technologies include the chemical synthesis of genes, the creation of recombinant DNA or recombinant RNA, cell fusion, plasmid transfer, transformation, transfection and transduction
Host	A cell whose metabolism is used for growth and reproduction of a virus, plasmid or other form of foreign DNA
Host-vector system	Compatible host/vector combinations that allow the stable introduction of foreign DNA into cells
Kudzu	A vine imported from Japan to the south eastern United States; an extremely prolific and difficult-to-control weed in its new environment
Microencapsulation	A method in which DNA is encapsulated in a substance, which allows the DNA to be taken up by cells
Microinjection	Injection of DNA into a cell or cell nucleus using a fine needle under a microscope
Microorganisms	Microscopic living entities, which can be viruses, bacteria or fungi
Mutagenesis (directed)	Selective alteration of a given gene using specific DNA cutting enzymes and mutagenic agents
Mutagenesis (undirected)	Random alteration of a gene or genetic sequence using mutagenic agents
Niche	A specific part of an ecosystem to which an organism has been able to adapt itself
Oncogene	A gene involved in the formation of cancers
Phylogenetic	Relating to evolutionary history and relationships between organisms based on that history
Plasmid	Small circular, self-replicating forms of DNA often used in recombinant DNA experiments as acceptors of foreign DNA
Plasmid transfer	The use of genetic or physical manipulation to introduce a foreign plasmid into a host cell
Procaryote	A cell or organism that lacks membrane-bound, structurally discrete nuclei and organelles. Procaryotes include bacteria and blue-green algae. (See eucaryote)
Recombinant DNA (rDNA)	The hybrid DNA resulting from joining pieces of DNA from different sources
RNA	Ribonucleic acid; found in three forms—messenger, transfer and ribosomal RNA; assists in translating the genetic code of a DNA sequence into its complementary protein
S. cerevisiae (*Sacromyces cerevisiae*)	Known generally as brewer's yeast, this organism is responsible for the fermentation of maltose to beer
Trait	A characteristic which is coded for in an organism's DNA.

Transduction	The transfer of one or more genes from one bacterium to another by a bacteriophage (a virus which infects bacteria)
Transfection	A process in which a bacterium is modified in a way which allows the cell to take up purified, intact viral or plasmid DNA
Transformation	The introduction of new genetic information into a cell using naked DNA, i.e. without using a vector
Vector	A transmission agent (i.e. a plasmid or virus) used to introduce foreign DNA into host cells
Virus	An infectious agent smaller than a bacterium that contains RNA or DNA as its genetic material and that requires a live host cell for replication
Wild-type (gene or organism)	The form commonly found in nature

SOCIAL SCIENCE LIBRARY

Manor Road Building
Manor Road
Oxford OX1 3UQ
Tel: (2)71093 (enquiries and renewals)
http://www.ssl.ox.ac.uk

This is a NORMAL LOAN item.

We will email you a reminder before this item is due.

Please see http://www.ssl.ox.ac.uk/lending.html
for details on:

- loan policies; these are also displayed on the notice boards and in our library guide.

- how to check when your books are due back.

- how to renew your books, including information on the maximum number of renewals. Items may be renewed if not reserved by another reader. Items must be renewed before the library closes on the due date.

- level of fines; fines are charged on overdue books.

Please note that this item may be recalled during Term.